REARTICULATIONS OF ORTHOPAEDIC SURGERY

THE PROCESS OF SPECIALTY BOUNDARY FORMATION
AND THE PROVISION OF FRACTURE CARE IN AMERICA

To gain an affiliation of this sort it does not suffice to imitate or translate or admire Poe; one must meet him, retrieve him, and inherit him in that place where the fate of the past has yet to be decided by the descendants.

Robert Pogue Harrison, *"The Dominion of the Dead"*

Care is accustomed to act, to take the initiative, to stake its claims, yet powerlessness and even helplessness are as intrinsic to the lived experience of care as the latter's irrepressible impulse to act, enable, nurse, and promote.

Robert Pogue Harrison, *"Gardens: An Essay on the Human Condition"*

REARTICULATIONS OF ORTHOPAEDIC SURGERY

THE PROCESS OF SPECIALTY BOUNDARY FORMATION
AND THE PROVISION OF FRACTURE CARE IN AMERICA

Kenneth Robert Gundle, MD

2014

First Printing: 2014

ISBN 978-1-312-53958-7

Self-published, with Lulu as distributor.
Contact the Author for questions on special ordering.

www.linkedin.com/in/kennethgundlemd/

Cover image: from Nicolas Andry, *Orthopédie*, 1741.

Contents

Preface & Acknowledgements

Like many preclinical students, I took advantage of numerous lunch and dinner talks during my first year at Harvard Medical School. These were an opportunity to learn about various fields, and had free food as an extra incentive to attend. I had an early interest in orthopaedic surgery, and eagerly went to a talk about the specialty by Dr. James Herndon. While finishing the presentation, Dr. Herndon mentioned that he was looking for a student to help research the history of orthopaedics at HMS. I thought it was a unique opportunity to explore my interest, while also continuing the sort of humanities scholarship that I enjoyed in college.

That was the beginning of an invaluable collaboration with Dr. Herndon and Dr. Harry Rubash at Massachusetts General Hospital. As I started researching the history of orthopaedics at Harvard, it fascinated me how much the field has expanded and transformed over a relatively brief period of time. While I knew that successful arthroplasty and arthroscopic surgery were technologies from the last fifty years, it was surprising that the treatment of fractures was predominately undertaken by general surgeons until World War II. This seemed a topic worth further study, and a meeting with Professor Allan Brandt spurred me to pursue it with more detailed research.

I was very fortunate to be introduced to Dr. Scott Podolsky, who agreed to advise me in shaping this project into a thesis. Dr. Podolsky's mentorship, as he pushed me to think deeply, read broadly, and write clearly, has helped me develop as a scholar and student of medicine. I am grateful for all the time he spent in discussions, in closely reading my drafts, and most of all in encouraging me throughout this process.

This work was made possible by the kind assistance of the dedicated and knowledgeable staff at the Countway Center for the History of Medicine. In particular, Jack Eckert and Jessica

Murphy never failed to help me track down a document, or give suggestions for additional materials. At the MGH Archives, Jeff Mifflin also supported this project through access and aid in finding annual reports and other sources.

Now, as a resident orthopaedic surgeon at the University of Washington, and with medicine changing all around us, I am thankful of the perspective gained through what was originally a medical school thesis project. In case others within or beyond orthopaedics may benefit, I have revised this work and made it available through modern self-publication – which itself is challenging professional boundaries. Please feel free to contact me with questions, insight, or critique. I write this not as a wise and experienced orthopaedist, nor as a historian, but as a student of both disciplines.

Additionally, I want to thank my family and friends – most of all my wife Megan. On the relatively rare nights, weekends and holidays that medical students and residents have free, I was often off reading and writing. Though this pursuit meant a bit less time together, you saw how important it was to me, and gave much appreciated, sustaining support.

Kenneth Gundle
September 2014

Abstract

The dramatic transformation of orthopaedics from the late 19th century through the end of World War II provides an opportunity to assess the process of professional boundary formation in a medical specialty. From its original meaning "to raise a child straight," orthopaedic surgery has iteratively rearticulated itself, and the path to its current authority in fracture care is not explained by a natural division of labor. Giving greater voice to individual practitioners, as they respond to changes in the social, political, and technological context of care delivery, contributes to our understanding of how divergent possibilities for demarcating specialty domains are negotiated. As orthopaedics and the healthcare profession face continuous pressures from internal and external forces, this narrative provides a lens through which to assess the imperative of adaptability and its consequences.

Introduction

Modern medicine is dominated by specialization, and subspecialization.[1] There are demarked boundaries based on organs, patient age, gender, disease, and whether treatment is surgical or medical. This fragmented system has implications for patient care, provider reimbursement, and medical education. How does a field define itself, and establish authority for the treatment of certain conditions? What forces influence the shaping of a specialty, and mediate the outcome when multiple groups lay claim on a set of patients? These are broad and important questions for understanding our medical system.

Studying the course of development of orthopaedic surgery is useful to assessing the question of specialization, and the process of professional boundary formation. Orthopaedics is a wide-ranging field, itself now subdivided, and reaching far beyond its original etymological meaning of "to raise a child straight."[2] In the period from the end of the 19th century through the 1950s, the scope of orthopaedic surgery enlarged and maneuvered dramatically. When it first arrived in the United States, pioneer orthopaedists primarily treated congenital conditions such as clubfoot, and were not involved in the acute care of fractures. By the end of World War II, however, the field had become preeminent in treating all traumatic injuries to the musculoskeletal system of the extremities and spine. By exploring how orthopaedic surgeons came to treat these fractures, it will become clear that the field iteratively redefined itself in relation to surgical methods, overarching principles of care, and particular anatomical areas. In response to changing social, economic, and

---

[1] Institute of Medicine. *A Manpower Policy for Primary Health Care.* (Washington, DC: National Academy of Sciences Press, 1978).

[2] Its ill-defined, somewhat nebulous name may allow greater latitude in shifting its meaning and focus. Lipscomb PR. Orthopedics, orthopaedic surgery, musculoskeletology, orthopaedics. *J Bone Joint Surg Am* 1975; 47:872-6.

technological developments, orthopaedic surgeons re-articulated the field's boundaries to further their position in the medical profession.

What has come before?

Several theories have been put forth to explain medical specialization. A prominent understanding maintains that ever-expanding medical knowledge and surgical technique naturally leads to practitioners caring for a smaller range of conditions with a more constrained set of treatment modalities. This process is often referred to as the natural division of labor. Especially to modern readers, in full knowledge of the exponential increases of knowledge, the inability of any one person to master even a significant portion of medicine is self-evident. In a history of American cardiology, Fye refers to this justification of specialization as the "conventional wisdom" and a pragmatic response to an increasing knowledge base resulting from discovery, invention, and experience.[3] What this theory fails to explain, however, is precisely where and how divisions are made. The multitude of specialties, based on various factors such as age, gender, organ system, or treatment modality, illustrate that many lines of demarcation are theoretically possible.

The natural division of labor theory fails to sufficiently explain the process of specialty definition, and redefinitions. Instead, we will see that a variety of social, economic, technological, and cultural frames have continuously shaped the emergence of orthopaedics, particularly how it relates to establishing authority over the treatment of fractures.[4] Urbanization, financial motivations, the transportation revolution, and war are part of the background and foreground in which orthopaedic surgery formed. Its practitioners

---

[3] Fye WB. *American cardiology: the history of a specialty and its college*. (The Johns Hopkins University Press. Baltimore, 1996) 2.

[4] Other recent works on medical specialization have argued along similar lines. Fye, *American cardiology*, takes this position in discussing cardiology. Regarding rehabilitation, which will be discussed in "From War to War", see: Gritzer G, Arluke A. *The making of rehabilitation: a political economy of medical specialization, 1890-1980*. (Berkeley: University of California Press, 1985).

responded to changing conditions, exploited opportunities, and moved through critical junctions that could have gone otherwise.

Prior histories of orthopaedic surgery have largely highlighted internal developments, with discoveries and discoverers as their focus.[5] The natural divisions of labor, and the beating drum of progress, are the lenses through which orthopaedics has been described. An exception is Cooter's *Surgery and Society in Peace and War*, which integrates the history of orthopaedics in the United Kingdom within a social, economic, and political framework.[6] He writes:

> *Essential components for the history of a medical specialty include concepts, pioneers, techniques, and institutions. These are necessary but not sufficient, as politics, economic interests, ideological shifts regarding childhood, and war have also significantly shaped the formation of modern orthopaedics.[7]*

In addition to focusing on the American experience, we will diverge from Cooter's analysis by developing how orthopaedists in each generation perceived their field in response to these internal and external forces. We will give a larger voice to the active participation of individual actors, who responded to divergent choices at given moments in history. Their active rearticulations of orthopaedics as it relates to other specialties adds to our understanding of professional boundary formation.

---

[5] Le Vay D. *The History of Orthopaedics*. (Park Ridge, NJ: Parthenon Publishing Group, 1990); Urbaniak JR. *A History of The American Orthopaedic Association: Leadership in Orthopaedic Surgery since 1887*. (American Orthopaedic Association, 2007); Rowe C. *Lest we forget: Orthopaedics at the Massachusetts General Hospital and War I, 1900-1996*. (Dublin, New Hampshire: William L. Bauhan, Publisher, 1996).

[6] Cooter R. *Surgery and Society in Peace and War: Orthopaedics and the Organization of Modern Medicine, 1880-1948*. (The Macmillian Press Ltd: London, 1993).

[7] Ibid 11.

Why write this?

This monograph presents a narrative combining the professional development of American orthopaedics, in particular its process of forming and altering the boundaries of the field, in relation to intertwined socioeconomic influences. We focus on the history of orthopaedics from its arrival in America through the end of World War II, with special attention to people and hospitals affiliated with Boston and Harvard Medical School. This narrows our lens, and takes advantage of the significant resources of the Countway Library of Medicine's Center for the History of Medicine, as well as archives at several Boston hospitals. In this period, the orthopaedic surgeons at Harvard-affiliated hospitals were often the leaders and drivers of the field's development. Their participation was critical to the growth and transformations of orthopaedics, and their work provides fundamental insights to this specialty's rise to a major force within the medical profession.

Following this introduction, "Setting the Stage" will focus on the origins of orthopaedics. Its initial conception was radically altered by the development of tenotomy - the cutting of tendon(s) to treat deformity - as a surgical technique. The pioneers of orthopaedics in Boston illustrate the tension between operative and mechanical treatments, at a time when changes in the landscape of medical care shifted to permit specialization. As the field began to organize in the late 1800s, there was vibrant debate as to whether surgery had a minor or major role in the field. We will see an orthopaedics initially confined largely to the treatment of chronic conditions of childhood, to the exclusion of acute fractures, with adults entering as a potential patient population at the opening of the 20th century.

"An Imperialistic Principle" shows how the young specialty actively sought to expand its boundaries during World War I. By articulating the field as focused on restoring and rehabilitating function to the locomotive apparatus, orthopaedists would successfully advocate for their role in treating acute battlefield injuries, including fractures. Serving the objectives of the military and government, appropriating an interest in the end results of treatment, and a willingness to rapidly train more practitioners, were critical to the transformation of the field during this conflict.

After the war, orthopaedists sought control over fractures and other injuries of industry, both due to financial opportunity and declining rates of their traditional domain of childhood conditions. "From War to War" discusses the inability of the field to gain immediate control over fracture care, and the disputes that arose with general surgeons. Detailing the development of the MGH Fracture Service grounds the analysis of this critical juncture, when orthopaedists responded to divergent ideas of authority over trauma by reconceiving their scope as surgery of the extremities and spine. Only in the course of World War II did orthopaedic surgery become predominant in the treatment of fractures, though this 'operative turn' opened the possibility of encroachment by the emerging field of physical medicine and rehabilitation.

These iterative reimaginings of orthopaedic surgery occurred at the convergence of societal demands with economic and technological circumstances, through the active participation of individual practitioners. The ability, or necessity, of a specialty to adapt its message and mission to a changing environment is a primary thesis of this work. In the final section, after discussing how this narrative adds to our understanding of the process of professional boundary formation, I will suggest how such a perspective aids in approaching modern dilemmas in orthopaedics and medicine more broadly.

The sources for this work have included primary literature such as hospital annual reports, archival documents, published articles, and presidential addresses from orthopaedic associations. Prior books written on American medicine, orthopaedics, and rehabilitation have aided in framing the broader discussion. Running notes in the body of this monograph will direct the interested reader to further information, and sources.

What does this contribute?

This work, I hope, adds to our knowledge of how one specialty, orthopaedic surgery, was conceived and developed during a tumultuous period. Its practitioners expanded and contracted the field's boundaries, postured themselves as an answer to the demands of society, and in the process redefined itself with each subsequent generation.

American medicine is facing significant changes, at the policy level as well as from new technologies, shifting dynamics in the patient-doctor relationship, and globalization. Work hour requirements are focusing attention on resident education,[8] and there are demands to simultaneously increase quality while justifying cost.[9] Individual specialties, as well as the entire medical community, will be required to reassess their scope and aims in response to changing conditions. Analyses such as the one presented here illustrate that iterative reimaginings of a field can dramatically change its course.

As orthopaedists openly wondered in the 1930s, much of American medicine is today asking whether it is "for permanence or for ending."[10] How we collectively and individually respond to our present conditions, along with a tincture of time, will determine the results.

---

[8] Institute of Medicine. *Resident duty hours: Enhancing sleep, supervision, and safety.* (Washington, DC: National Academy of Sciences Press; 2008).

[9] Institute of Medicine. *Crossing the Quality Chasm: A New Health System for the 21st Century.* (Washington, DC: National Academy of Sciences Press; 2001).

[10] Goldthwait JE. The backgrounds and foregrounds of orthopaedics. *J Bone Joint Surg Am* 1933; 15:279.

Setting the Stage
Emergence & Disputed Ground of Orthopaedics

Introduction

Here we will describe the state of orthopaedic surgery from its origins until World War 1. At the field's conception, mechanically preventing and treating 'deformities' of children was considered the essential mission. The emergence of tenotomy as an operative technique dramatically altered the practice of orthopaedics, as did the development of anaesthesia and antisepsis that enhanced the opportunities for surgical treatment.

Closely examining the lives and practices of some of the first Boston physicians to consider themselves orthopedic surgeons will clarify the conditions they saw and their efforts at treatment in the 19th century. A changing ecology of care in response to rising population density, transportation opportunities, and increasingly hospital-based medical care, facilitated would-be orthopaedists in specializing as the century drew to a close. In this period, orthopaedic surgeons both directed and benefitted from a charity-based medicine, which brought attention and philanthropic funding necessary to their treatment of crippled children.

Growth in Boston orthopaedics was mirrored elsewhere, and at a time when various segments of society were organizing, likewise the American Orthopedic Association (AOA) was founded. This organization, and the journal it promulgated, was a means to begin defining orthopaedics and increase its legitimate recognition by other physicians and the public. As the number of orthopaedists began to grow and organize, debates began about the relative place of operative and mechanical treatment in the specialty. The prominent conception held that the field should limit its work to chronic conditions. Acute fracture care, with few exceptions, was the accepted purview of general surgery.

The Confusing Word

The underlying meaning of 'orthopaedic surgeon' is not immediately understood by the public or even large segments of the medical profession. When asked to describe an orthopaedic surgeon, the likely response is a general idea of what one does, or the sort of conditions one treats, or even some stereotyped attributes. How this field is defined, or more specifically how it continues to be redefined, is a central task of this work. It is fitting to start at the term's recognized origins.

In 1741 a book was published entitled, "Orthopaedia: Or, the Art of Correcting and Preventing Deformities in Children."[1] The author was Dr. Nicolas Andry (1658-1742), a Professor of Medicine in the Royal College and senior Dean of the Faculty of Physick in Paris. Etymologically, the author meant to combine the Greek words for 'straight' and 'child' to create orthopaedics, with his book then being an encyclopedia on the topic.

While the term has survived, the intended audience of the book, and its content, denote a meaning far removed from modern orthopaedics. For his audience, Dr. Andry was quite specific in the book's second subtitle: "By such Means as may easily be put in practice by Parents themselves, and all such as are employed in Educating Children." This was not meant as a textbook for practitioners, but rather as a home and school guide to raising children without deformity.

---

[1] Andry N. *Orthopaedia: or, the art of correcting and preventing deformities in children.* Translated from French. In two volumes. (London: A. Millar, 1743).

But what did Andry mean by deformity? The book begins with descriptions of various parts, and then turns to means of prevention and treatment for various conditions. The pathologies mentioned range from spinal curvature to aberrations from the 'normal' face and eyebrows, and included excessive sweating of the hands, poor complexion, and even freckles. The best form for function is repeatedly stressed by Andry to be the best form aesthetically; much of the book deals with means to prevent or correct seemingly undesirable physical characteristics. Fractures are only discussed to say that their treatment is difficult, that a physician and surgeon ought be consulted, but that a severe deformity is likely to result.[2]

In order to prevent and treat these conditions, special attention is given to proper shoes and chairs. Appropriate posture while studying, sewing, sleeping and standing are prescribed to preserve 'straight' bodies in children. Andry wrote, "Of all the Methods proper for preserving Health, and for preventing, and even curing a great number of Diseases, there is none equal to moderate Exercise."[3] He detailed the treatment for rickets:

> *Would you prevent or cure the Rickets? What is properer for this, than to put the Child who is threatened with that Disease into a Swing, fastened about this Breast and Neck, and so moved backwards and forwards as often as you please? For by this Motion frequently repeated, the Ligaments of the Joints are relaxed and distended, and this is considerably affected by those Springs which the Child gives for Joy of being thus swung; for this Spine, Arms and Legs are by these Boundings put upon the stretch.*[4]

---

[2] Ibid v1, 130.

[3] Ibid v2, 212.

[4] Ibid v2, 214.

Function, pathogenesis, and treatment are all related to underlying mechanics in the paradigm proposed by Andry. Future orthopedists have read into Andry's text to find evidence that the principle of obtaining the best function of the body is the ideal of the field.[5] Several have appreciated his focus on prevention as fodder for the need to bolster this part of their practice. It is less common to discuss Andry's concern with freckles, hand sweating, and other conditions and treatments, many of which now fall into the domain of dermatology or plastic surgery, if considered pathologic at all.

While the mechanical, functional focus on diseases and treatments was passed down along with the name, even those following soon after Andry did not faithfully carry on with his concerns. The focus on children, the core of his book and part of the field's etymological basis, would fade away gradually in the 20th century. This name, its spelling, and its usefulness or appropriateness would continue to be a telling dispute on the unsettled boundaries of the profession.[6]

Towards Surgery:
club foot, tenotomy, and orthopaedics in Europe

One condition discussed by Andry was congenital equinovarus (club foot). The standard treatment of the time, which consisted of various methods and forms of serial wrapping of the affected foot to attempt gradual correction, is mentioned in the Hippocratic cannon. In Europe during the century after Andry's *Orthopaedia*, both physicians and so-called irregular practitioners like bonesetters cared for children with club foot. Attempts to surgically treat this condition introduced the technique of tenotomy, and the widespread

---

[5] Goldthwait JE. The backgrounds and foregrounds of orthopaedics. *J Bone Joint Surg Am*. 1933; 15:279-301.

[6] Lipscomb PR. Orthopedics, orthopaedic surgery, musculoskeletology, orthopaedics. *J Bone Joint Surg Am* 1975; 47:872-6.

adoption of this procedure significantly contributed to making orthopedics a surgical field.

William John Little (1810-1894) would become England's strongest advocate for tenotomy: the surgical treatment of club foot and other conditions by cutting one or more tendons subcutaneously.[7] His purpose was personal, as he himself had a malformed foot in childhood that developed into an equinovarus position. Studying medicine at the London Hospital in 1826, Little became acquainted with the work of a French surgeon, Jacques Matheiu Delpech (1777-1832), who had cut the Achilles tendon to aid in treating club foot at least once. He asked Delpech to operate on his foot, but was advised against it due to the risk of infection; this was in a time prior to sterile technique and long before antibiotics, when every surgery bore a significant chance of sepsis.

Passed up for a surgical staff position, Little out of necessity decided to become a medical doctor. Doing so required study abroad, and Little travelled to Berlin University, where he continued dissection-based studies on deformed feet. While in Berlin he heard of a surgeon in Hanover, Louis Stromeyer (1804-1876), who had modified Delpech's approach to treat club foot by cutting the Achilles tendon subcutaneously.

Stromeyer, the son of a surgeon, did not come upon the decision to cut the Achilles lightly. According to Sir Arthur Keith's *Menders of the Maimed*, Stromeyer accepted for treatment the 14 year-old son of a local schoolmaster who had severe, painful club foot.[8] While difficult to believe by modern length-of-stay standards, Stromeyer spent eighteen months treating this case as an inpatient. When he finally did cut the tendon, he did so by a small incision, going just

---

[7] Le Vay D. *The History of Orthopaedics.* (Park Ridge, NJ: Partenon Publishing Group, 1990) 499.

[8] Keith A. *Menders of the maimed: the anatomical & physiological principles underlying the treatment of injuries to muscles, nerves, bones, & joints.* (J.B. Lippincott Company: Philadelphia, 1919) 68.

subcutaneously to reach the Achilles. After the procedure he could dorsiflex the patient's foot, and gradually the equinovarus was improved and the tendon healed.

In Keith's telling, Stromeyer immediately recognized the possibility of tenotomy to more effectively treat numerous types of deformities that had previously been construed as outside the reach of surgeons. He applied it to strabismus by cutting ocular muscles, and considered its use in scoliosis. To Stromeyer it was contracted, pathologic muscle action that permitted the misshaped bones, and tenotomy could provide release. And, to use the modern phase, his 'minimally-invasive' approach of subcutaneous incision decreased the risk of infection.

Little proceeded to Hanover and was treated successfully by Stromeyer. After his own operation Little stayed on, learning the technique himself. Keith writes, "He returned to Berlin cured and convinced that a new era had dawned for the deformed."[9] Back in England, in 1838, he founded a charity hospital for the maimed poor, the Orthopaedic Institution (later renamed the Royal Orthopaedic Hospital). He lectured to students on orthopaedic treatment of deformities, and wrote several books, including *The Nature and Treatment of the Deformities of the Human Frame.*[10]

Although a physician who was overlooked for a surgical position, Little advocated tenotomy under the name of orthopaedics and focused on the treatment of deformities. The wide range of Andry's *Orthopaedia* was shifted by the advent of a new procedure, tenotomy, that pushed treatment of conditions like club foot away from parents or bonesetters and towards surgeons. Stromeyer and Little are well remembered in the history of orthopaedic surgery. They pushed forward an operative correction of a condition that had

---

[9] Ibid 70.

[10] Little WJ. *The nature and treatment of deformities of the human frame.* (London, 1853).

long been treated by mechanical techniques. By doing so the standard orthopaedic treatments of buckles and straps gained a surgical element, marking a critical juncture in the history of the infantile specialty.

Pioneers in Boston:
orthopaedic practice in 19th century America

The opportunity provided by tenotomy spurred develops in orthopaedics in Europe as well as America. In Boston, it was Dr. John Ball Brown (1784-1862) who took up this procedure and started a special institute for treating orthopaedic conditions. The clinical practices of John Ball, as well as his son Buckminster Brown (1819-1891), illustrate the era's scope of care as seen by early Harvard-affiliated orthopaedic surgeons. Patients came to them with chronic conditions, and received a combination of tenotomy supplemented with mechanical treatment. The experience of this father and son also demonstrate the path to success in this young field in mid-19th century America.

John Ball Brown, following a preceptorship and a degree from Harvard Medical School in 1813, was appointed a surgeon to the Boston Alms House, a government-sponsored facility to provide care for the poor.[11] While organized in 1811, the Massachusetts General Hospital did not open until 1821, after which time Brown was made an associate surgeon, later to be consulting surgeon.[12] He was a general surgeon, without evidence of orthopaedic interest in his early practice.

---

[11] Shands AR, Jr. *The Early Orthopaedic Surgeons of America.* (Saint Louis: The C.V. Mosby Company, 1970) 129.

[12] He would marry a daughter of John Collins Warren, a founder of MGH and prominent surgeon whose supported Brown's career.

It was his experience as a father that drew him to problems of the spine and other deforming conditions. His first son passed away after 'inflammation' of the spinal cord, which was a description for spinal tuberculosis. He second son, Buckminster, had debilitating scoliosis. These intimate exposures to the problems of spinal deformity encouraged John Ball to action, as he writes in an published letter, "... my attention has been forcibly drawn to the study and treatment of spinal diseases generally, and to the correction of other deformities of the human body, such as distortion of the limbs, club feet, etc., etc."[13]

The timing of his personal interest aligned with Little's publication of his experience with subcutaneous tenotomy. To further his cause, in 1838 Brown founded the Boston Orthopedic Institution, located at 65 Belknap Street (now Joy Street), located off Beacon Street and the Boston Common in Beacon Hill, just two blocks from the MGH.[14] Its announcement in *The Boston Medical and Surgical Journal* illustrate its novelty by a description of the roots of orthopaedics and detailing its objective to treat "spinal distortions, club-feet, etc.," particularly in children.[15] Public understanding of its name may have been a problem, evidenced by its otherwise being referred to as the Orthopedique Infirmary, or later, receiving the subtitle of Hospital for the Cure of Deformities of the Human Frame.

The Institution could hold at least 57 patients, and had rooms for exercises and apparatus. Tenotomy was, however, central to the practice of this new infirmary. John Brown performed one of (if not the) first tenotomy for club foot in America in 1839, following closely Little's report of its use in London in 1837. According to a history of the Boston Orthopedic Institution, "Brown popularized this procedure and performed as many as one hundred and ten

---

[13] Cohen J. The Boston Orthopedic Institution. *J Bone Joint Surg Am* 1958; 40:1176-1178.

[14] Ibid.

[15] Editorial: Orthopedic Infirmary. *Boston Med and Surg J* 1839; 19:369.

tenotomies in two years, most of them for foot deformities, but some for spinal curvatures as well."[16] Manipulations and devices to hold the body in certain positions supplemented the operations.

John Ball Brown, and later his son Buckminster, presented a variety of cases they treated at the Massachusetts Medical Society's Annual Meetings, and later published five full case series. Patients with club foot, lateral spine curvature, cervical vertebrae caries, torticollis, and hip disease are described. Quoting a few of these in full will show their methods of treatment and approach to these patients:

> *Case I:*

> *Talipes. -- The first case to which I will draw attention is the one from which this case and this photograph were taken... A boy born with such a distortion of the leg and foot that the great toe was turned up against the side of the knee, and, when the child was awake, was in close contact with the internal condyle of the femur. The tibialis anticus and posticus muscles were strongly contracted - structurally shortened. The treatment consisted in the division of the tendons of these muscles, and in the use of a vairety of apparatus, employing sometimes the spring and sometimes the screw power. By these means the leg and foot were gradually brought into a normal position. By this time, however, this result was somewhat more than half accomplished, the tendons, growing more rapidly than the bones, had united, and again presented an obstacle to further improvement. These were re-divided, and, in about six months, the result was as shown in the second cast. The second photograph gives a correct idea of the foot when it was nearly straight. When I last saw the child, he walked on the sole of his foot...*

---

[16] Cohen, *The Boston Orthopedic Institution*, 1177.

Case XX:

The last case, of which I have to speak this morning, is that of girl twenty years of age, whose situation before treatment is shown in this photography. When eleven years old, while at play, she was thrown form a height of sixteen feet, by the caving in of an embankment, and lumbar and sacral spine striking upon a large stone. The fall produced insensibility for a few moments. She then recovered and went to school. She continued her usual avocations for five or six weeks, growing, daily, more and more feeble. She was then attacked with agonizing pain in the lumbar region, followed by complete loss of sensation and motion below the hips. The thighs and legs gradually contracted, until the left knee was forced against, and under, the right thigh, and the thigh was drawn up to an acute angle with the body, and twisted to the right. These parts were in such close contact that it was with difficulty I forced the knee from under the thigh where it had lain for years. The patient had extreme lateral curvature, with excessive incurvation of the lumbar vertebrae. The first photograph was taken nine years after the accident. By means of subcutaneous division of tendons in the groins, popliteal regions and in the feet, followed by mechanical appliances, together with a carefully adjusted spinal apparatus, the girl was in three months straightened out as seen in the second representation.[17]

These two cases are representative of the pathology and therapeutic strategy of the Institute. A combination of surgical tendon release and gradual mechanical treatments using a variety of devices to hold parts of the body purposefully in position; length of treatment was long, to the order of months. Most patients were children, in keeping with Andry, but nowhere do we see an interest in complexion, hand sweating, or other more aesthetic concerns of the *Orthopaedia.*

---

[17] Brown B. *Cases in orthopedic surgery: read before the Massachusetts Medical Society at its annual meeting, June 3, 1868.* (Boston: David Clapp & Son, 1869).

Although an initial injury may have been sustained, they presented to the Institute in a chronic state.

A prominent New York orthopaedist and surgeon to the Hospital for the Ruptured and Crippled,[18] Virgil Gibney (1847-1927), gave a definition of what constituted the scope of his field while speaking at a state medical society in 1880. It closely mirrors the conditions described in Brown's cases:

> *By orthopedic cases is meant diseases of the joints and bones in their immediate vicinity; lateral deviations and rotations of the spinal column; muscular contractions from diseases with affect their nerve supply or the tissues themselves mechanically; the various forms of paralysis par excellence, essential, spinal, or infantile; and the congenital deformities, including talipes equino varus. The above enumeration fairly covers the diseases that fall under the care of the orthopedist; and it will be observed that the majority are essentially chronic.*[19]

John B. Brown, Gibney, and others who focused their career on treating deformities saw their work as chronic care. Another prominent New York orthopaedist of some note stressed that his field has no role in acute care: "Orthopaedic surgery lies wholly within the domain of 'chronic' surgery."[20] Fracture care, in addition to being neglected in their case logs, is expressly excluded from orthopaedics by many American pioneers.

---

[18] This institution was later renamed the Hospital for Special Surgery. 'Ruptured' as used here refers to hernia, cases of which were a significant part of the hospital's early work.

[19] Gibney VP. *Notes on the management of orthopedic cases, reprinted from The Medical Herald.* (Louisville, KY: John P. Morton & Co., Printers, 1880) 1.

[20] Shaffer NM. *What is orthopaedic surgery?* (New York: G.P. Putnam's Sons, 1890) 18.

Favorable preconditions for specialization:
A changing ecology of care

In the second half of the 19th century, aspiring orthopaedic surgeons in Boston and elsewhere began finding a foothold. John B. Brown started as a general surgeon, working in a broad practice for decades. Family illness, and the possibilities afforded by tenotomy, propelled him to focus on 'orthopaedy.' Within eight years of work at the Orthopedic Institution, Brown was arguing for specialization:

> *The practice of orthopaedy is a distinct branch of surgery as much as dentistry or the profession of the oculist or aurist and should be practiced exclusively. It certainly requires all of any one man's mind to treat these deformities judiciously. It would be better for the profession and for the public at large if the duties of the profession were more divided and subdivided; for the same reason that the mechanic arts are carried to a higher state of perfection in proportion as their distinct branches are made the exclusive objects of attention by individuals.*[21]

He is arguing for a natural division of labor, resulting from increased knowledge necessary for adequately treating these conditions. However, there were social, economic, and broader technological developments that facilitated the increasing success at specialization in orthopaedics and other fields. This favorable milieu enabled orthopaedists by name and practice to emerge.

Until well into the middle of the 19th century, economic conditions in rural America encouraged most families to care for themselves. In Starr's *The Social Transformation of American Medicine*, he writes, "Dispersed in a heavily rural society, lacking modern transportation, the great majority of the population was effectively cut off from

---

[21] Brown JB. *Report of cases in the Boston Orthopedic Institution, or Hospital for the Cure of Deformities of the Human Frame.* (Boston: D. Clapp, Jr., 1844).

ordinary recourse to physicians because of the prohibitive opportunity cost of travel."[22] The geography of rural life meant that the cost of medical care included not only the direct services of treatment, but also the indirect cost of physician travel and the time of whoever must go find him.

For the physician, reaching patients meant riding along country roads and carrying whatever supplies might be necessary. Railroads, and subsequently automobiles, revolutionized travel for both doctor and patient. Among the first to buy cars, editorials by physicians with titles like "Would not practice without an auto" began appearing in the *Journal of the American Medical Association*.[23] Patients could come from further afield, and consultants could travel by rail across larger distances. Harvard orthopaedists would soon be traveling throughout New England to run clinics for poliomyelitis.

While increased ease of travel and communication were helpful, a rise in population density was critical. Increasing urbanization and rising population density in cities like Boston brought sufficient 'clinical material' - as annual bulletins at Harvard Medical School and elsewhere refers to patient population[24] - to allow physicians to focus their work and still see enough patients to make a living.

Urbanization and industrial wage earning jobs also brought to the cities a higher proportion of people who were unmarried, often separated from their families, and had little recourse when sick but to visit a hospital. In the early 19th century, Starr writes, "Almost no one who had a choice sought hospital care."[25] Patients were

22 Starr P. The Social Transformation of American Medicine. (USA: Basic Books, 1982) 68.

23 Kessel G. Would not practice without an auto. *JAMA* 1908; 50:814.

24 Emphasis on the rich 'clinical material' available at MGH and other Harvard affiliated hospitals was prominently mentioned in the HMS bulletins. Harvard Medical School, in: A Catalogue of the Officers and Students of Harvard University for the Academical Year 1891-1892. (Boston: University Press, 1892).

25 Starr, *The Social Transformation of American Medicine*, 72.

generally seamen, travelers, the poor, and the solitary aged. However, industrialization and wage jobs meant less family members to care for the sick at home. Growing urban populations lead to more lodging houses, laundries, tailors and restaurants. Likewise, Vogel argues in a book on healthcare of this period in Boston, the hospital was another such 'corollary' institution.[26] By then end of the 19th century, the hospital was becoming the central location for medical care.[27]

The clinic or hospital also became the location for particular equipment, which physicians could not carry or afford individually. Brown argued for specialization in orthopaedics along these lines:

> *Deformities of the human frame cannot be conveniently and judiciously treated except in a hospital or institution expressly devoted to this object. It is not for the interest of any general practitioner of medicine and surgery to be at the expense of furnishing himself with the variety of apparatus (some of which is very expensive), required in treating these deformities. Then, again, he might not always have at command suitable mechanics. I have been obliged to keep two or three first-rate mechanics constantly in employ for several years in making and altering and repairing apparatus... The Orthopaedic Institution at Paris, under the direction of Dr. J. Guérin, has nearly $100,000 invested in apparatus. The Boston Institution has a considerable amount and is constantly increasing.[28]*

The existence of orthopaedic apparatus itself became a reason for specialized orthopaedists. Similar arguments would be made in other emerging fields, like the electrocardiogram for cardiologists.[29]

26 Vogel MJ. *The invention of the modern hospital, Boston, 1870-1930.* (Chicago: University of Chicago Press, 1980).

27 Rosenberg CE. *The care of strangers: the rise of America's hospital system.* (New York: Basic Books, 1987).

28 Brown, *Report of cases in the Boston Orthopedic Institution.*

29 Fye WB. *American cardiology: the history of a specialty and its college.* (The Johns Hopkins University Press: Baltimore, 1996).

The entry cost dissuades other competitors, patients see equipment particular for their care, and a clinic with the proper equipment narrows its practice in line with their use.

The natural growth model suggests that increasing complexity and technological developments drive specialization and the rational division of labor. The situation in American medicine and orthopaedics in the late 19th century suggests otherwise. It was in large cities that orthopaedic hospitals and clinics first appeared and survived as part of broader transitions in the localization of medical care. Specialization was made possible by a convergence of altered social conditions.

The ability to begin specializing, however, does not explain the scope of practice that orthopaedic surgeons claimed. Specialization is a necessary but not sufficient condition for professional boundary formation within medicine. Why did early orthopaedists focus on the crippled child? How did the field's practitioners begin to organize, define themselves, and advocate for a privileged position? The opportunities for specialization that came with urbanization and industrialization were only the starting point for orthopaedic surgery as a formed entity.

Prominence and Expansion:
Orthopaedics at Boston Children's Hospital and
Massachusetts General Hospital, 1869 - 1910

The development of orthopaedics in two Harvard institutions, the Children's Hospital and Massachusetts General Hospital, illustrate a transition from a successful focus on children to a broader view of the field that included adults. Benefiting from an environment conducive to specialization, orthopaedists at these hospitals solidified their place in the medical community and began to expand their reach. Understanding their clinical practice prior to World War

I sets the stage for analyzing how the field would redefine itself during the conflict, and afterwards.

Charity, Children, and Orthopaedics

The bread and butter of the first generation of American orthopaedists was treating 'crippled children' with apparatus and tenotomy. The focus on children garnered the support of philanthropic groups, wealthy patrons, and the government during a charitable era of medicine. Here we see a first attempt to carve out a special place for their field in the provision of organized medical care.

Roger Cooter, in discussing the development of orthopaedics in pre-World War I Britain, argues that it is no coincidence that early orthopaedists first focused on crippled children.[30] The welfare of children, including their health, was a significant target of governmental and non-governmental organizations. Industrialization brought the potential to put young children to work at wage earning jobs, while philanthropic groups and government responded with compulsory education and child labor laws. Likewise, deficient sanitation and healthcare for young, poor, and often immigrant children became a strong appeal for an active Christian charity.

In Boston, the founding of the Children's Hospital was a beneficiary of this philanthropic focus. Opening to patients in July 1869, its stated purpose was as a charity to serve "the sick poor of Boston between the ages of two and twelve."[31] The first board president was Nathaniel Thayer, a wealthy businessman and noted philanthropist.[32]

30 Cooter R. *Surgery and Society in Peace and War: Orthopaedics and the Organization of Modern Medicine, 1880-1948*. (London: The Macmillian Press Ltd, 1993) 53-78.

31 Smith CA. *The Children's Hospital of Boston: built better than they knew*. (Boston: Little Brown and Company, 1983) 2.

Upon its founding, its nursing staff was a voluntary staff of nuns, and creating a 'moral' environment was part of the treatment ethos.

As a specialized pediatric hospital, the institution was also responding to the increased population density and other societal changes discussed previously. While the MGH would admit some children, the founders of Children's argued for particular treatment and study of the diseases of childhood. The first patient was a seven-year-old girl with a fractured radius, the second an eight-year-old boy with a fractured femur; auspicious for orthopaedics, except that these children were treated by general surgeons.[33] The first orthopaedist at Children's, Dr. Edward H. Bradford (1848-1926), joined in 1878.

Bradford was a descent of a Massachusetts Governor, who graduated Harvard College and entered HMS in 1869 - the year Children's Hospital opened, as well as the beginning of Charles W. Eliot's term as President of Harvard University. While at HMS, Bradford experienced the significant reforms that Eliot enacted. After serving as a House Officer at MGH he spent two years studying in Europe. His interest in orthopaedics was spurred by work with Buckminster Brown, and time subsequently spent in New York studying under eminent orthopaedists.[34] At the Children's Hospital, Bradford and the many students he would train benefited from this charity-supported hospital, where they could exclusively practice in their field.

Another factor that spurred on orthopaedics at the hospital was the prevalence of skeletal tuberculosis in the pediatric population. Out

---

32 Ibid 26.

33 One of the first general surgeons at Children's, Dr. Arthur Tracy Cabot, would later leave to be a urology specialist at MGH, but not before inventing a wire splint for leg fractures that was among those officially adopted for use in WW1. Ibid 70.

34 Shands AR, Jr, *The Early Orthopaedic Surgeons of America*, 140-56.

of 862 patients admitted at Children's between 1883 and 1885, 214 had skeletal tuberculosis.[35] At the time, the standard in treatment for these cases was prolonged skeletal traction. The frame of iron pipes and canvas, devised by Bradford and known by his name, were a familiar sight on the wards. Other congenital deformities admitted for treatment included club foot, dislocation of the hip, scoliosis and kyphosis, as well as rickets. Its surgical appliance shop, necessary for making frames, braces, traction devices and splints, preceded the hospital's first laboratory by a quarter century. A history of the hospital notes, "... for most of its first fifty years, the Children's was primarily known as an orthopedic hospital, and a good one."[36] Given the prevalence of disease, we can understand why Bradford was considered the lead surgeon at Children's from 1883 onwards. Here, at least, orthopaedists had found prominence not at their own special institute, but at a pediatric hospital.

This position brought accolades, for Bradford and orthopaedics. He received the first orthopaedics appointment at Harvard Medical School, as Clinical Instructor in Orthopaedic Surgery, in 1880.

35 Smith, *The Children's Hospital of Boston*, 65.

36 Ibid 67.

37 Bradford EH, Lovett RW. A treatise on orthopedic surgery. (New York: William Wood & Co., 1890).

38 Shands AR, Jr, The Early Orthopaedic Surgeons of America, 140-56.

39 The orthopaedists at Children's embraced their charitable mission, and were concerned with the social conditions of their patients. Heliotherapy, the practice of putting children out in the sun for specified periods of time, became a well-valued adjuvant treatment for skeletal tuberculosis. In search of fresh air and sunshine, a convalescent home in the countryside was established, and future wards were constructed to allow for adequate sun exposure. Perhaps some had intuitively determined the positive benefits of sunlight on bone formation, before the identification of Vitamin D. There was also a concern for making these children self-sufficient. Bradford considered his most important work to be the Industrial School for Crippled and Deformed Children, founded in 1893. The goal was to educate handicapped children and train them for jobs in industrial America, so that they could earn a living independent of private or public charity. In her book In Sickness and in Wealth, Rosemary argues that up into in the early 1900s, hospitals retained principles of "scientific charity" as articulated after the Civil War, with a preference for providing self-help, delivered by private organizations acting in a paternalistic manner to improve society. Stevens R. In Sickness and in wealth: American hospitals in the twentieth century. (New York: Basic Books, 1989) 19.

Bradford was influential in founding the American Orthopedic Association, to be discussed below, and served as its president in 1889. He trained Dr. Robert B. Lovett (1859-1924), and together they wrote *Orthopedic Surgery*, first published in 1890, which went through several editions and was long considered the field's standard textbook.[37] The book is focused, understandably but meaningfully, on conditions of childhood. With both local and national prominence, Bradford rose to Assistant Professor of Orthopedics in 1893 and in 1903 became the first full Professor of Orthopaedic Surgery at HMS, and later served as Dean of the Medical School.[38]

Children in Boston with physical disabilities, as well as the orthopaedists who took a leading role in their medical treatment and social reintegration, [39] benefited from philanthropists who took a special interest in this issue. Conditions like club foot, scoliosis, developmental dislocations of the hip, and skeletal tuberculosis were prevalent in this age group and became the clinical focus for Bradford and other orthopaedists. The charitable support for crippled children and their medical care, as exemplified by Boston Children's Hospital, encouraged a focus on this population.

Slow Beginning to Orthopaedics at MGH

It is a fitting contrast that an orthopaedics department at MGH developed slowly. Certainly there were adults with severe contractures and other conditions that caused pain and limited their function, but they were not early beneficiaries of special philanthropy; they did not have their Tiny Tim. It was not until 1900 that the position of Consulting Orthopaedic Surgeon was created and filled by Dr. Joel E. Goldthwait (1866-1961).[40]

---

40 Eighty-seventh Annual Report of the Trustees of the Massachusetts General Hospital 1900. (Boston: The Barta Press, 1901) 8.

MGH was not entirely unfavorable towards specialists. For years there had been special physicians to out-patients with skin disease, nervous system disease, and throat disease, as well as an ophthalmic surgeon, an aural surgeon, and a pathologist on staff. Rather, several developments in the provision of care generally, and in orthopaedics specifically, facilitated its eventual growth. Firstly, the expanding out-patient clinics at MGH allowed additional patients to be seen without requiring occupancy of the limited beds. This is particularly important for mechanical treatment, which was often prolonged. Secondly, the possibility of more rapidly treating deformities by surgery meant a decreased length of stay for those who might be admitted. Although anesthesia was displayed fifty years prior at MGH, the growing acceptance and use of antisepsis and asepsis in the 1890s permitted surgery not only to be painless, but more likely beneficial to patients. "Asepsis has been perhaps the most powerful aid on the surgical side in facilitating the efficient handling of so many additional patients," wrote a resident physician in 1898.[41]

It its annual report for 1900, a total of 28 operations tabulated as 'orthopedic' are listed.[42] However, the largest percentage of these was for correcting deformities of the nose (eleven);[43] there were seven tendon repairs and four tenotomies, though two of these were for strabismus. An operation for hallux valgus and hammer toe, as well as breaking adhesions around the elbow, finish the list. These represent what, at least according to the tabulated records of the hospital, what was considered orthopaedics in the same year that Goldthwait became the field's consultant at MGH. Deformities of the nose were included, as with Andry, and the use of tenotomy for strabismus likely led to its sustained inclusion. In contrast, under the general surgical category there were 22 leg amputations, 19 arm

---

41 Eighty-fifth Annual Report of the Trustees of the Massachusetts General Hospital 1898. (Boston: The Barta Press, 1899) 45.

42 Eighty-seventh Annual Report of the Trustees of the Massachusetts General Hospital 1900

43 It is unclear whether it was orthopaedists themselves who operated for deviated septum, and the records do not clarify who performed the operations for joint tuberculosis. Also, the classification system itself was a work in progress. Record keeping in this annual reports was progressing on a yearly basis.

amputations, and 12 fractured patella that were wired, not to mention 311 operations for appendicitis and 178 surgical repairs of hernia. These statistics provide insight as to what constituted orthopaedics, and what was considered beyond its scope, during the field's first years at MGH.

Though on the Board of Consultation, Goldthwait was not added on the staff until 1903.[44] This was the year that an orthopaedic department was established, and the following year an out-patient clinic was created. Goldthwait's title became Surgeon to the Orthopedic Out-Patient Department. The amount of designated staff grew; as of 1905 Dr. Robert B. Osgood was appointed Assistant Orthopedic Surgeon to Out-Patients, with four additional assistants to the orthopedic surgeons. One of these assistants, Dr. Max Böhm, was also named Surgeon in Charge of the Medico-Mechanical Room.[45] An expanding demand for their services is evidenced by an annual report notice that the hospital shop had to be enlarged in order to produce and maintain the necessary equipment, and that a ward had been renovated for use by the department.[46]

The 1905 annual report provides additional details of care in by the expanding out-patient departments.[47] That year, a total of 21,874 new patients presented to a clinic, with the orthopaedics department seeing 1,026 new patients and 4,126 established patients. Individual diagnoses are also reported, with the number of patients that were seen by each department. Some conditions were predominately seen by orthopaedists, like hallux valgus (29 orthopaedics vs 4 surgery), genu varus or valgum (12 vs 2), sacroiliac disease (56 vs 1), and

---

44 Ninetieth Annual Report of the Trustees of the Massachusetts General Hospital 1903. (Boston: The Barta Press, 1904) 17.

45 Ninety-first Annual Report of the Trustees of the Massachusetts General Hospital 1904. (Boston: The Barta Press, 1905).

46 Ibid 5, 46.

47 Ninety-second Annual Report of the Trustees of the Massachusetts General Hospital 1905. (Boston: The Barta Press, 1906).

arthritis (more than a 100 seen by orthopaedics). Diseases largely treated by general surgery included dislocation (8 orthopaedics, 53 surgery), fractures (14 orthopaedics, 719 surgery), osteomyelitis (5 vs 22), and sprains (12 vs more than 200). These numbers illustrate a degree of overlap, but also clear trends. Orthopaedics had found a niche treating chronic deformities, but acute conditions like fractures and dislocations were largely seen by the general surgery staff. Diseases requiring major operations, like sarcomas, were also treated by general surgeons (14 vs 1). Overall, only 5% of the new patients seen in the various out-patient clinics went to the orthopaedics department, and less than 10% of these orthopaedics patients received an operation.

While most of these conditions were similar to those found in children, the treatment of arthritis was a significant expansion. In the following years there was much speculation about the etiologies of various forms of arthritis, with many believing that all were caused by infection.[48] The classification of rheumatic, infectious, and osteoarthritis was debated, with atrophic and hypertrophic forms being the preferred descriptions in many papers.[49] Successful arthroplasty was decades away; most treatment involved rest and then regaining motion. Excision of joints with a goal of creating a painless fusion was occasionally done. This sort of radical operation appears at MGH to have been performed largely by general surgeons, and we will see that forthcoming debates about the boundaries of the field centered around this kind of new, more invasive operations and whether orthopaedists should attempt them.

Orthopaedics as Mechanical, Physical Therapy

---

48 Osgood RB, Soutter R, Bucholz H, Danforth MS, Low HC. Report of progress in orthopedic surgery. *Boston Medical and Surgical Journal* 1913; 168(13):461-7.

49 Osgood RB, Soutter R, Bucholz H, Danforth MS. Third report in orthopedic surgery, reprinted from the *Boston Medical and Surgical Journal*. (Boston: W.M. Leonard, 1913).

Primarily it was mechanical therapy, under the direction of orthopaedists, that grew rapidly as a treatment modality for patients with a broad variety of conditions throughout the hospital and its clinics. A special device called the Zander apparatus, imported from Sweden, was installed in 1904 in large spaces both at MGH and at the McLean Hospital. Their function is detailed in the annual report:

> *They include pieces for the active exercises of the arms and chest, for the trunk, and for the legs; the others are for passive exercises of various parts of the body, and are those which seemed best fitted for the needs of our patients. An electric--motor furnishes the power, the shafting being boxed in along the wall for concealment and protection.*[50]

Both these Zander devices, and the surgical apparatus shop, were praised as financially self-supporting and a distinctly valuable advantage to the Orthopedic Department and the hospital at large. "The Medico-Mechanical Department," a trustee report noted, "has proved to be a valuable adjunct to the Hospital. It is still under the efficient direction of Dr. Max Böhm."[51] Essentially, orthopaedists were directing what would now be considered physical therapy, with additional use of braces and other apparatus. The Zander machine itself served more than ten thousand patients in 1905, in addition to the orthopaedic clinic. The following year hydrotherapeutic apparatus were installed; even more momentous, ground was broken on a new Orthopedic Ward at MGH.[52] Opening on November 6th, 1907, it added 18 beds to the hospital and included areas for women and men, as well as children, representing a structural marker of orthopaedics transcending the crippled child. In celebration of Ether Day that year, the new Orthopedic Ward was opened to public tours.[53]

---

50 Ninety-second Annual Report of the Trustees of the Massachusetts General Hospital 1905, 213.

51 Ibid 44.

52 Ninety-third Annual Report of the Trustees of the Massachusetts General Hospital 1906. (Boston: The Barta Press, 1906) 47.

53 Ibid.

Transition:
Conflict, as Orthopaedics Approaches Adults & Operations

By the numbers, orthopaedic surgery at MGH as it first developed was overwhelmingly involved in non-operative therapy. Surgeries were increasingly performed, but at levels far below the surgical department and for a narrow set of conditions. But their work was valued, and rewarded with clinical space, staff positions, and financial support for equipment. Moreover, in the period of a few years orthopedics had opened itself up to the care of adults. Dr. Brackett, who joined of MGH orthopaedics in 1907 and became its chief, spoke of this shift during his term as President of the American Orthopaedic Association:

> *It is within the remembrance of even most of the younger men, when the orthopedic work was almost entirely confined to children. Now a large, if not the larger, part is devoted to adults, and in the work with them here has come relief to one of the most helpless forms of cripples.*[54]

Orthopaedic surgery's first identity was intimately wrapped up with the treatment of crippled children. Expansion into the treatment of adults, and departments forming at large hospitals like MGH, brought the opportunity for conflict. When a Boston orthopaedist who worked with the Harvard group was offered the opportunity to start an orthopaedic surgery department at Boston City Hospital, he was irritated to be told by general surgeons that his work would be confined to treating clubfeet, scoliosis, and rickets.[55]

Negotiations about the scope of orthopaedics inside the Harvard community were mirrored elsewhere, and became an openly hostile debate within the growing group of physicians identifying

---

54 Brackett EG. President's annual address. *American Journal of Orthopedic Surgery* 1905; 3(1):1-5.

55 Mayer L. Reflections on some interesting personalities in orthopaedic surgery during the first quarter of the century. *J Bone Joint Surg Am* 1955; 37:384.

themselves as orthopaedic surgeons. The field's initial success, and its organization, provided an opportunity but also a necessity of professional differentiation.

Medicine, and Orthopaedics, Organizing

Hastened communication and transportation, along with an expanded industrialized marketplace, brought patients and doctors closer together. The location of care was shifting to hospitals and clinics. These developments also opened the possibility for increased competition, locally for individual physicians as well as for the profession more broadly. At the same time that labor unions and corporations began organizing on regional and national scales, the medical community was achieving greater unity. Starr points to the 1901 formation of the American Medical Association House of Delegates, set up as a confederation of state medical societies, as a turning point for the growth of the state societies and the AMA as a whole.[56] In the decade from 1900 to 1910, the AMA membership rose nearly ten-fold to seventy thousand, comprising half of all American physicians. A far smaller orthopaedic community formed its first national organization in this period, and for similar reasons.

In January of 1887 two New York orthopaedists, Drs. Virgil Gibney and Newton Shaffer, gathered a group of fourteen men to consider forming an association for their field. While two in attendence were opposed, and two more abstained, ten voted in favor and the American Orthopaedic Association (AOA) was born. The stated purpose was "the advancement of orthopaedic science and art."[57] Gibney was made the first chairman, while Harvard's Robert Lovett was made acting secretary. Thirty-five men were selected to comprise the original membership, and its first meeting took place that June. One of the first acts, arranged by Dr. Bradford, was to take

56 Starr, *The Social Transformation of American Medicine*, 109.

57 Mayer L. Orthopaedic surgery in the United States of America. *J Bone Joint Surg Br* 32(4):461.

all papers presented at the meetings for the use of the AOA to produce a journal entitled *The Transactions of the American Orthopaedic Association.*[58]

For an association like the AOA to gain power and become a voice for the specialty, individuals must be willing to transfer some authority to the group.[59] For the medical community as a whole, changes in the ecology of care were already promoting cooperation. The increase in legitimate medical complexity provided a common foundation for discussion, while increasing specialization and referrals within hospital-based care delivery generated a web of reliance among physicians. Doctors also depended on colleagues to defend them in medical malpractice cases that were becoming more frequent. This already complex system of care may have facilitated a willingness to concede some autonomy to a larger organization that could then represent and advance their interests as a whole. The AOA annual meetings, and the published *Transactions* that resulted, had formative and normative functions for the young specialty.

Looking back in 1905, an AOA president wrote that the *Transactions* was gaining in circulation, "... and by this wider distribution of the Transactions of our meetings the influence of the Association in advancing orthopedic work is more widely felt."[60] Prior to its publication, literature on orthopaedics had to be found in foreign or domestic general medical publications, or else in individually published monographs. The *Transactions* were a tangible statement of the field's separate existence, with the AOA as a driving force behind its progress. For aspiring orthopaedists, speaking at AOA meetings, publishing papers in its journal, and eventually obtaining membership became a path to prominence in the field.

---

58 Straub LR. Orthopaedics in a changing world. *J Bone Joint Surg Am* 1968;50:1037-42.

59 Starr, *The Social Transformation of American Medicine*, 111.

60 Brackett, President's annual address, 1.

Through the AOA, orthopaedics was forming an identity as a professional organization. Rosen, in book on specialization in medicine, describes several attributes of such a group.[61] A number of individuals must self-identify with the profession, and call themselves by its name. These individuals must be part of a community, with regular meetings. The AOA encouraged this identification as an orthopaedist, as only those with a special interest in the field and a track record of several year's work (often proven by regular presentations at their meetings) would qualify someone for membership. And, importantly, Rosen also argues that there must be a shared knowledge base.[62] The *Transactions*, which were rapidly increasing in shear quantity of work being published, were building that foundation.[63] While orthopaedics was taking on attributes of a professional group, this by itself does not explain how the spectrum of orthopaedic care was determined.

With each issue, the *Transactions* were becoming a shared knowledge base. What was published, additionally, served as an indicator of what orthopaedists considered their field of work. A publications committee oversaw the selection of papers and the process of choosing what to present at the annual meeting. By answering the question of what to accept and publish they were implicitly defining the field.

The AOA President in 1897 wrote a summary of the first ten years of papers in the *Transactions*.[64] The dominant focus was tuberculosis of the bones and joints, accounting for 114 of the 292 papers. Spinal tuberculosis (Pott's disease) by itself was the subject of 49 articles. The next most common subject was club foot, with 43 papers. Nearly absent is any mention of fractures. The contents of the first decade of

---

61 Rosen G. The specialization of medicine with particular reference to ophthalmology. (New York: Arno Press, 1972).

62 Ibid.

63 Brackett, President's annual address, 2.

64 Ketch S. The work and influence of the AOA. *Trans Am Orthop Assn* 1897; 10:1-7.

Transactions are a testament to the period's view of orthopaedics, and are consistent with the scope of care described at the Boston Children's Hospital.

More explicitly, editorials in the *Transactions* sought to increase standardization and unity in orthopaedics. Each year, the AOA President gave an address at the annual meeting, which was subsequently published as the opening paper of the *Transactions*. It was customary to discuss the progress that had been made in orthopaedic surgery, and its future direction.[65] For example, a President's address entitled "Standardization in orthopedic instruction and practice" encouraged increased standards in training and treatment, to be achieved by additional standing committees of the AOA.[66] These Presidential Addresses taken together are a yearly pulse of orthopaedics, and the efforts by its leaders to direct it.

Forming the AOA, however, did not bring unity to the practice of orthopaedics. While early by-laws required that its members focus their practice on orthopaedics for a period of time before being eligible for nomination to join, the obvious question became: what is orthopaedics? Simply publishing the clinical and research work of its members was not enough. The founding of the AOA occurs in continuity with a multitude of published lectures and monographs on this topic. Its organization forced the issue of defining the field; its annual meetings and publications became a forum for debating the scope and direction of orthopaedics. AOA Presidential Addresses were a cornerstone of these debates. Successive presidents and other authors attempted to influence the activities of anyone who would call himself an orthopaedist.

---

65 This tradition is mentioned in: Brackett, President's annual address, 1.

66 Freiberg AH. President's address – standardization in orthopedic instruction and practice. *American Journal of Orthoapedic Surgery* 1911; 9:1-5.

The major debate of the period was regarding the relative prominence of surgical operations versus mechanical therapy in the field. Another layer in the discussion was how broadly or circumscribed to define orthopaedics. The arguments suggest an unsettled field, with different conceptions of the present and future possibilities.

Orthopaedic surgery or Mechanical Orthopedy?

As discussed earlier, the effects of anesthesia and antisepsis were multiple for orthopaedics. Anesthesia meant that patients might submit to operation in situations other than absolute necessity. Lister's techniques gave surgeons confidence that wounds would heal, and that elective operations may proceed without sepsis as an inevitable result.

An active proponent for operative orthopaedics was Dr. A.M. Phelps of New York. As president of the AOA, he gave an address at the 1894 annual meeting.[67] It embraced a transition from 'orthopedy,' by which he meant treating deformities by braces and splints, to orthopedic surgery - the operative treatment of deformities by surgeons focusing in that area. He compared orthopaedics to ophthalmology and gynecology, which had already found success as surgical specialties. And for the domain of his field, he envisioned a broad grasp: Phelps was among the first to advocate that orthopaedic surgeons should treat all fractures and dislocations. He did not want to lose mechanical treatment, however, especially when arguing that hernias should also be treated by his specialty. He writes, "Hernia unquestionably should be classified as an orthopedic subject. It is as important and as difficult to adjust a support to remedy a hernia as a splint to hip-joint disease, and frequently, by mechanical means, herniae are cured."[68] Competition with general

---

67 Phelps AM. The influence of surgical bacteriology and modern pathology upon orthopaedic surgery, and past, present and future of that specialty. *Trans Am Orthop Assn* 1895; 7:31-42.

surgeons was, to Phelps, unavoidable. His rhetoric illustrates a perception that orthopaedists were subservient to their general surgery colleagues:

> *The orthopaedist was always at war with the general surgeon. There never was a time when they could lie peacefully together in the same bed, excepting like the lion and the lamb - one inside the other, and the poor orthopede was always inside.*[69]

Only by open efforts to claim fracture care (let alone hernias) away from general surgeons, Phelps argues, could their status rise in the hospitals. It is no coincidence that, apart from appendicitis, fractures and hernias were among the most common conditions seen by surgical clinics. Even subsuming a portion would significantly increase the number of orthopaedic patients. Avoiding a clear definition of the bounds of orthopaedics, especially as to where it ends and general surgery begins, Phelps' vision was of expansion, operation, and competition.

A major voice in opposition to an operative paradigm was Dr. Newton Shaffer, leader of the New York Orthopedic Dispensary, a founder of the AOA, and its only member to serve twice as president. He was an orthopaedist who entered the operating room mostly to apply plaster-of-Paris dressings, rarely even performing a subcutaneous tenotomy.[70] Yet he was clearly eminent in the field, further evidenced by his selection to speak at the Tenth International Medical Congress, held in Berlin. It had required some lobbying to secure the inclusion of an orthopaedic surgery section, with some arguing that these deformities are treated by irregular practitioners, and others unable to see the line separating the orthopaedist from the general surgeon. He addressed these concerns during his lecture, which was aptly titled "What is orthopaedic surgery?"[71] The lecture

68 Ibid.

69 Ibid.

70 Mayer, Reflections on some interesting personalities in orthopaedic surgery.

serves as a manifesto for the mechanically oriented orthopaedist, and included several arguments for professional differentiation along specific lines.

Early in the address, he confronted the controversy of his field's special inclusion at the conference: "Where shall the line be drawn? What is orthopaedic surgery? Shall orthopaedic surgeons be general surgeons as well, and shall general surgeons be orthopaedists? If these questions are answered in the affirmative, there is no room for a special orthopaedic section in the Berlin Congress."[72] His greatest concern was that mingling orthopaedics with general surgery would engender his field's existence.

Shaffer criticizes a perceived trend to surgical treatment. Bradford and Lovett's book, *Orthopaedic Surgery*, is seen as part of an invasion of general surgery by their inclusion of laminectomy, osteotomy, and osteoclasis. To him, the knife, saw, chisel and osteoclast are tools of the general surgeon, and a "stumbling-block" and diversion from genuine orthopaedic work. He recalls a conversation with a general surgeon:

> *The remark of a prominent general surgeon to [Shaffer], after reading the latest work on orthopaedic surgery, is not, perhaps, so much out of place. He said: 'The next work on orthopaedic surgery will likely tell us all about fractures and dislocations.'*[73]

Whereas Phelps and others would have seen this as an accomplishment, Shaffer saw a dangerous dilution of orthopaedics. The total lack of definitions of orthopaedics in other authors' works

---

71 Shaffer NM. *What is orthopaedic surgery?* (G.P. Putnam's Sons: New York, 1890). This was read before the orthopaedic section of the Tenth International Medical Congress, Berlin, August 5, 1890.

72 Ibid 9.

73 Ibid 14.

---

bothers Shaffer. He wanted to see the field formally defined, and provided his own:

> *Orthopaedic surgery is that department of surgery which includes the prevention, the mechanical treatment, and the operative treatment, of chronic or progressive deformities, for the proper treatment of which special forms of appartus or special mechanical dressings are necessary.*[74]

There are two important boundaries to orthopaedics, as delineated here. Firstly, and as noted before, this conceives of an entirely chronic scope; by definition, acute fractures and dislocations would be excluded. Secondly, mechanical treatment is a necessary precondition for orthopaedic care. If a simple operation is all that is needed, without apparatus afterwards, Shaffer advocated transferring that patient to a general surgeon. An orthopaedist excising a joint because he happens to be mechanically treating the patient's arthritis or other condition, was to Shaffer like a gynecologist operating for cataract if the patient happens to be a woman.[75] Mechanico-therapeutics is the heart of his orthopaedics. Surgical treatment is not excluded, but is subjugated. Shaffer argued, "Operative surgery has its own place, and in orthopaedic work that place should be second."[76]

Part of the motivation for defining the field as such was to avoid competition with general surgeons. In an editorial entitled "The relation of orthopaedic surgery to general surgery," published in the *Boston Medical and Surgical Journal* in 1891, Shaffer explained his conception of the appropriate rise of specialties:

---

74 Ibid 12.

75 Shaffer NM. Modern orthopaedic surgery - a general reply to criticism. *American Medico-Surgical Bulletin*, 15 January 1895.

76 Shaffer, What is orthopaedic surgery? 14.

> *The function of the orthopaedic surgeon should therefore be to fill a place not occupied by the general surgeon - to do a work that the general surgeon is either unwilling or unfitted to undertake, and to aid in developing an important department of surgery which has been too long neglected or ignored.*[77]

He is describing how the field first formed, by focusing on club feet and skeletal tuberculosis when it was perceived as neglected. He considered the very reason for the field's existence to be due to prior neglect of mechanico-therapy by the medical profession.[78] Doing that work "...without interfering in the slightest degree with the overcrowded ranks of the general surgeon" should be the goal.[79] Shaffer fails to recognize, however, that Brown and other pioneers aimed to displace non-physician practitioners like bonesetters or independent mechanics. Competition outside the boundaries of medical doctors is reasonable, but not within the profession as a whole. In this way, Shaffer is aligning himself with medicine and surgery rather than with orthopaedics.[80]

The debate over surgical versus mechanical orthopaedics was personal, at times resembling a public feud, with ad hominem attacks accompanying the rhetoric on both sides.[81] Professional differentiation is a political process, as social, historical, and

---

77 Shaffer NM. The relation of orthopaedic surgery to general surgery. Reprinted from the Boston Medical and Surgical Journal of February 26, 1891. (Boston: Damrell & Upham, 1891) 5.

78 Shaffer NM. On the definition and the scope of orthopaedic surgery: Remarks on Dr. Gibney's paper. Reprinted from The New York Medical Journal of November 14, 1891. (New York: D. Appleton and Company, 1891).

79 Shaffer, The relation of orthopaedic surgery to general surgery, 4.

80 A skeptic looking at Shaffer's non-operative stance might be tempted to attribute it to his hospital's lack of funds for a modern operating room - a problem he admits (Shaffer, Modern orthopedic surgery). New understanding in bacteriology, pathology, and asepsis seemed to preclude operating in ward rooms, as had been previously done. Shaffer may have been reflecting a reality - that for many orthopaedists in American a suitable operating room was simply not affordable.

81 Shaffer NM. Modern orthopedic surgery: a reply to Dr. Phelps. American Medico-Surgical Bulletin, 1 Jul 1895.

economic conditions are negotiated by particular historical figures. This runs counter to the natural division of labor theory, whereby increasing complexity and knowledge simply engenders specialization as a matter of course.

The contrast was not only operative versus non-operative treatment; all recognized the place for both, and in comparison to today the treatments undertaken were predominantly conservative. All sides were for growth, as we have already seen orthopaedics expand into treating adults. It was more a debate about the process of defining the specialty. Shaffer's definition was akin to a terminal differentiation, laying out sharp boundaries and principles, avoiding overlap or competition with other fields. He thought that success and survival of the field would be served by focusing on the mechanical therapy known especially to them, and walling that off from other physicians. It is also a static definition, suggesting that the reasons for the field's development must always be the purpose for its continuation. The other position, personified by Phelps, was of a pluripotent bent; it left room to expand and adapt, avoiding in particular marking out areas that are off-limits. The domain of orthopaedics to Phelps seems more like a view from atop a tower, with a clear grasp of its nearby bread-and-butter deformities but peering off at the more distant fractures and other acute conditions, hoping to bring them under its control.

Lacking resolution, moving forwards

There is no clear end to the operative versus mechanical therapy debate. Both surgical operations and mechanico-therapy grew, as we have seen at MGH. A shift towards increasing variety of operations by orthopaedists did continue. Shaffer was replaced at the New York Orthopedic Dispensary in 1898 by Russell Hibbs, who was surgically-oriented.[82] In 1911, Hibbs was one of two orthopaedic

---

82 Mayer, Orthopaedic surgery in the United States of America.

surgeons who independently published a method to fuse the spine for tubercular disease using bone grafting. They sought to replace plaster corsets, frames, and braces with an internal bony splint.[83] While conservative therapy that included heliotherapy was the standard treatment for skeletal tuberculosis in children, these advances in bone grafting were significant for surgical treatment for a variety of conditions. Shaffer's worry, that physical and mechanical therapies for deformities would be pushed aside by an operative focus, did not materialize until the 1950s, when physical medicine and rehabilitation was able to wrest some control over this arena.[84]

More so than ongoing developments in methods of treatment, it is notable that no circumscribed definition of orthopaedics emerged. Its imprecise name, whose etymologic roots at already been cast aside along with much of the content as imagined by Andry, was an advantage suited to the situation in the late 1890s and early 1900s. The AOA President in 1896 recognized this in his address. Since the field had already left its roots of treating only children, he concluded, "One is obliged, therefore, to define not what orthopedy means, but what its present application has come to be."[85] Its imprecision, far from a problem, is argued as a potential boon:

> *... the specialty differs from others that are limited to special organs or tissues in that it has a broader outlook and an ill-defined boundary; but this is not a disadvantage, since the object of a specialty is not the complete separation of one branch of medicine from another, but rather to provide the opportunity for the concentration of energy in a particular direction... An ill-defined boundary is rather in advantage, if thus, by more*

---

83 Hibbs method utilized small chips of bone, while Frederick H. Albee favored contoured pieces of tibial graft, which he harvested and crafted using a motor saw. The two were rivals. Albee had been a surgical house pupil at MGH in 1903, when Goldthwait was surgeon to the orthopedic out patient department. Albee would later state: "Never train around a disability that can be removed."

84 Further discussion of the loss of rehabilitation, see Part Four.

85 Whitman R. A definition of the scope of orthopaedic surgery as indicated by its origin, by its development and by the work of the AOA. Trans Am Orthop Assn 1896; 9:1.

---

frequent contact with workers in other fields, we may escape the accusation of narrowness that is sometimes urged against special work...[86]

This statement well describes how orthopaedics proceeded up until World War 1. Its boundaries were porous, but it had a consistent set of core conditions and methods. While visions for the future of the field existed, the reality in Boston was an orthopaedics focused on a few childhood conditions, a growing interest in adults, and only an insignificant role in acute fracture treatment.

Perhaps this reflected an inability of the AOA leadership to dictate its membership's practice. Individual surgeons, even if they considered their clinical work to be confined to orthopaedics, nonetheless had practical constraints in the degree to which specialization was possible, or desirable. Also, constraining its reach would run counter to the goal of spreading orthopaedics by increasing the number of practitioners and their positions in hospitals and medical schools throughout the country. In 1905, based on a certain degree of success in pursuing the AOA goal of furthering the field, its president recommended relaxing membership fees and requirements, so that the AOA badge of membership might give aid to a young surgeon eager to start a community's first orthopaedic clinic or hospital appointment.[87] This sort of expansionistic posture will play a central role in the events of World War I.

---

86 Ibid.

87 Brackett, President's annual address.

An Imperialistic Principle:
Expanding Orthopaedic Borders in the Great War

Introduction

If we consider modern orthopaedic practice, and its dominance of musculoskeletal trauma to the extremities and spine, it would seen a foregone conclusion that orthopaedic surgeons would be prominently involved in World War I. Yet, we have seen, the treatment of fractures was held securely within the domain of general surgery. Orthopaedic practice in early 20th century America largely consisted of treating children with bone tuberculosis, poliomyelitis, or clubfoot, though interest in adult deformity was growing. An orthopedist's surgical forte was primarily limited to tenotomy, and possibly osteotomy or tendon transplantation.

It was not inevitable that this group of surgeons would enter the war and take center stage in directing treatment of the wounded. Its own membership initially doubted their usefulness at a war hospital. And when they did go to Europe, it was not for the purpose of treating fractures. It was the potential to prevent, treat, and restore injured soldiers to battle or industry that drove orthopaedic involvement early in the war.

By the end of the war, American orthopedic surgery had expanded its clinical reach, the number of its practitioners, and achieved increased recognition as a specialty. Dr. Joel E. Goldthwait, an HMS graduate and one-time chief of orthopedic surgery at MGH discussed earlier, who himself would play a prominent role in these accomplishments, sounded a hopeful note for the future:

The position which orthopedic surgery has been forced to

occupy in this war is so conspicuous and so much of the work that is now being developed is bound to be preserved, to meet the civil needs when the war is over, that is not out of place to consider some of the features which have led to this position and that will demand its permanence.[1]

The purpose of this chapter is to assess how orthopaedists expanded the boundaries of their field and gained a foothold in the care of acute musculoskeletal injuries. First, an overview of the events of the war will be presented. From this background will follow an analysis of the internal and external forces that shaped orthopaedics in this era. During the war, orthopaedic surgeons reconceived their specialty as committed to the principle of maximizing locomotive function. Applying this principle to fracture care, the field successfully argued to professional, military, and political authorities that orthopaedic surgeons would serve the interests of the armed forces and the treasury, as well as soldier-patients. This shift built off emerging attempts to assess the end results of surgical treatment. A systematic, coordinated delivery of care from initial injury through rehabilitation was advocated. In addition to a principle-based orthopaedics, the imperialistic posture of the field aided attempts to expand their borders to include acute injuries, as evidenced by large-scale efforts to train new orthopaedists at home and abroad.

The scope of orthopaedics was widened by this principle-based definition coupled with an expansionistic outlook. The question of the field's permanence would figure prominently once hostilities ended and orthopaedics was forced again to redefine itself. Wartime gains would not be easily or quickly transitioned to civil life.

Overview of Events

---

[1] Goldthwait JE. The place of orthopedic surgery in the treatment of war casualties. *The Military Surgeon* 1917; 41:450-6.

A Doubtful Beginning

Compared to prior conflicts, the outbreak of World War I brought a different spectrum of pathology. High velocity machine-gun bullets, explosives and accompanying shrapnel at the close range of trenches combined with the bacteria-laden, fertilized soil to generate significant morbidity. Early in the war, an open femur fracture resulted in 80% mortality.[2] The scale of the war, the technology employed, and the environmental milieu resulted in masses of wounded, infected soldiers.

Before United States entry in the war, groups of Americans physicians went to Europe to aid in treatment. Among those who went were Harvard orthopaedic surgeons Robert B. Osgood (1873-1956) and Nathaniel Allison (1876-1932). For 3 months in the spring of 1915, Osgood served as orthopaedic surgeon in the Harvard Medical School Unit at the American Ambulance in Paris. Upon his return, Osgood published and lectured widely on his experience. He had initially questioned the usefulness of this endeavor:

> We confess to some doubt at the beginning, as to whether an orthopaedic service at a war hospital would prove interesting to the orthopaedic surgeons or be useful to the other departments of the hospital. The doubt as to the interest of the orthopaedic problems presented disappeared at once and the work assigned to the orthopaedic surgeon, seemed to answer the other question.[3]

In a separate forum, he repeated this sentiment: "At the onset of the war many of us were uncertain whether orthopedic surgery had any important service to offer."[4] This initial doubt is telling. It was by no

[2] Ellis H. *A history of surgery.* (London: Greenwich Medical Media, 2001) 136-8.

[3] Osgood RB. Orthopaedic work in a war hospital. *Boston Medical and Surgical Journal* 1916; 174(4):109-27.

means ordained that these orthopedists would be needed in the wartime division of labor. The experience Osgood had, and how he became interested and useful, is therefore important to understanding the subsequent course of development.

Some of the work referred to the orthopaedic surgeon was familiar. Immediately he was given cases pertaining to joint pathology, ranging from penetrating trauma involving a joint to any ankylosis secondary to immobilization or inflammation. Physiotherapy installations were initiated to begin range of motion and massage treatment. A nearby convalescent home was established to continue therapy and watch over ambulatory cases.

Nerve injuries that resulted in paralysis and the need for apparatus were also treated by the orthopaedist, as were tendon contractions. Osgood mentions that lessons learned from poliomyelitis could be applied to the care of these injuries. For example, well known "cock-up" wrist splints were used to treat wrist drop following damage to the musculospiral (radial) nerve, and wire ankle-foot splints were given to soldiers with foot drop from damage to a peroneal nerve.

Injuries to the 'soft parts' presented another opportunity to contribute. "Not a few internal derangements of joints occurred as a result of the wrenches and falls which are invited by trench and camp life. A slipping or fracture semilunar cartilage, a ruptured crucial ligament, a torn supraspinatus tendon, may be a truly crippling injury if it is neglected, but is usually completely amenable to appropriate treatment."[5] And once treated, the soldier was ready to return to active duty; a point that, we will see, was certainly not lost on the writer or audience.

---

[4] Osgood RB. The orthopedic phases of military surgery. *Transactions of the College of Physicians of Philadelphia* 1917; 39:97-103.

[5] Ibid 98.

There were also fractures, both simple and compound (i.e., open). Fractures involving the head were handled by another team, possibly by Dr. Harvey Cushing, who spent time there serving as a director of the Harvard Unit. In three months Osgood cared for 19 patients with simple fractures, and 99 cases of open fractures of the extremities. These difficult cases drew Osgood's attention, as he reported,

> *The compound fractures of the shafts of the long bones represented, however, the most interesting group of cases, since they presented problems which were to us new, and their solutions seemed to be accomplished by somewhat unanticipated measures. They were practically all septic and this sepsis was in the majority of cases severe and accompanied by marked rise in temperature and constitutional reaction.*[6]

Unlike the stiff joints and need for apparatus, these acute fractures presented novel challenges. Osgood finishes his report by encouraging further work in the war, writing, "It seems to us most desirable that American orthopaedic surgery should accept this opportunity for service, which the unfortunate war has made necessary."[7]

Preparations and American entry into the War

In 1916, following Osgood's experience in Europe, the Orthopedic Section of the AMA appointed a committee to assess orthopaedic preparedness. A quarter century after the American Orthopaedic Association was founded, an AMA section for the specialty had been created in 1912.[8] This AMA acceptance of orthopaedics as a specialty

---

[6] Osgood, Orthopaedic work in a war hospital, 114.

[7] Ibid 127.

[8] In contrast, physiatry had yet to be recognized. Orthopaedics did not yet face this competition for claims to the work of rehabilitation. The encroachment of rehabilitation by physiatry is discussed further in "From War to War".

gave the field standing at this meeting, and an opportunity to advocate for forming this committee. A similar committee of five men was created at the 1916 AOA meeting. To facilitate their work, Goldthwait was made chairman of both. The purpose for these groups was to assess the likely orthopaedic needs, to determine what apparatus might be best applied in the setting of war, the orthopaedic hospital equipment that should be standard, and how best to secure a sufficient number of orthopaedic surgeons.[9] Goldthwait also opened communications with the Surgeon General and other national authorities as the country as a whole was preparing for hostilities.

In April 1917 the United States entered the war. The first British requests were for six base hospitals, as well as twenty orthopaedic surgeons to assist Sir Robert Jones (1857-1933).[10] That the initial cable specifically identified orthopaedic surgeons as a particular need bolstered their position in the American Army.[11] Prior relationships between Jones and American orthopaedists, facilitated by the formation of the AOA, its published journal, and trips to study in Liverpool by a generation of practitioners including Goldthwait, had paid dividends.[12] Within a month Goldthwait collaborated with the Surgeon General, organized a unit of orthopaedists, and set sail for England. None in this group were servicemen, and all quickly made preparations to leave their private practices.

---

[9] Goldthwait JE. *The division of orthopaedic surgery in the AEF*. (Norwood, MA: Plimpton Press, 1941) 4.

[10] The storied career of Robert Jones will be mentioned at times, but a full recounting of his life is outside the scope of this paper. Numerous articles on his contributions, as well as that of his uncle H.O. Thomas, exist in the literature. One such example is: Hagy M. 'Keeping up with the Joneses.' *The Iowa Orthopaedic Journal* 24:133-137.

[11] Ibid 10.

[12] Ibid 6. The relationship between American and British orthopaedists is further detailed in: Mayer L. Reflections on some interesting personalities in orthopaedic surgery during the first quarter of the century. *J Bone Joint Surg Am* 1955;37:374-83.

Over the ensuing summer, the AOA committee was instructed to prepare a guide for orthopedic practice in the military. Also, a member of their preparedness committee was asked to join the Surgeon General's office. The Surgeon General created a Division of Orthopedic Surgery within the Army's Medical Department. Dr. Elliot Brackett, a Harvard trained orthopedist then working at MGH, was named to lead this division. His first tasks were to obtain the necessary number of adequately trained surgeons, and to organize plans for reconstruction of the wounded in the United States.

Structure of Wartime Orthopaedics upon American Arrival

Goldthwait led twenty orthopedists to England in May 1917, with the majority being from Boston or Boston-trained.[13] This journey to Europe was not unlike the trips abroad taken by earlier generations of surgeons such as Buckminster Brown or John Collins Warren. Though with the additional purpose of military service, they had similar goals: exposure to a wealth of clinical material, and interactions with highly regarded practitioners like Robert Jones to learn from.

By the Harvard orthopedists' arrival in 1917, 7850 orthopaedic beds had been set aside within general military hospitals in Great Britain, with additional centers being built and further demand apparent.[14] Large orthopedic centers included Leeds, Alder Hey near Liverpool, and Shephard's Bush in London. Osgood describes the Old Mill Military Hospital in Aberdeen as being well equipped, containing a gymnasium, curative workshops, hydrotherapy and electro-therapy, operating rooms, a photographic studio, and an enlisted artist to make sketches of important cases.

---

[13] Osgood RB. The orthopedic centers of Great Britain and their American medical officers. *The American Journal of Orthopedic Surgery* 1918; 16(2):132-40.

[14] Ibid.

War provided a bureaucratic and hierarchical environment more structured than civil practice. Who would treat what, in war, could be ordered and made a matter of policy. In England, Robert Jones had been appointed inspector of military orthopedics the year prior. An official classification of conditions considered orthopedic had been created:

> (a) Derangements and disabilities of joints, simple and grave, including anchylosis. (b) Deformities and disabilities of feet, such as hallux valgus, hallux rigidus, hammer toes, metatarsalgia, painful heels, flat and claw feet. (c) Malunited and ununited fractures. (d) Injuries to ligaments, muscles, and tendons. (e) Cases requiring tendon transplantation or other treatment for irreparable destruction of nerves. (f) Nerve injuries complicated by fractures or stiffness of joint. (g) Cases requiring surgical appliances.[15]

This was the partition of care that existed when American orthopedists arrived. Neither simple nor open fractures, in the acute phase, were within their purview except when involving joints. Jones had been helping his uncle H.O. Thomas reduce fractures since a small boy, and treated them as part of his pre-war practice, yet these were left to the general surgeons under this directive. Malunions and nonunions, as chronic deformities, were notably included as orthopaedic. In this framework, fractures were initially treated elsewhere, but if deformity results then care was to be transferred to an orthopaedist. This was in keeping with arguments that the proper scope of orthopaedics included only chronic ailments.[16]

---

[15] Goldthwait JE, The place of orthopedic surgery in the treatment of war casualties, 451.

[16] The arguments for including only chronic cases within orthopaedics, which were put forth in the decades prior to WW1, were discussed in "Setting the Stage."

[17] Shattuck GC. Medical work in the British Armies in France. *The Military Surgeon* 1919; 45:248-56.

The orthopaedic centers in England, where arriving American surgeons began working, were part of a larger system of care. The wartime structure of medical care helps in understanding some of the developments in treatment. Dr. George Shattuck published a detailed description of the work with the British armies in France while stationed with the Harvard Unit at No. 22 General Hospital.[17] The continuum for care of an injured soldier involved several layers, with triage occurring at each location. Field ambulances, Casualty Clearing Stations, and general hospitals were major organizational components.

A field ambulance was attached to every fighting division. Its purpose was to collect the wounded, perform initial stabilizing measures, and evacuate as quickly as possible those needing additional care. Stretcher-bearers were needed in large number, and the principle interventions included placing splints and dressings to stop hemorrhage and facilitate transport by ambulance cars. At this stage only provisional diagnoses were made, mostly to facilitate timely triage. Abbreviations flourished, with examples including PUO for 'pyrexia of unknown origin,' and NYD for 'not yet diagnosed.'

Casualty Clearing Stations (CCS) received wounded from the Field Ambulances, made more definitive diagnoses, and performed a great deal of the operating. While originally designed to be mobile, the emergence of trench warfare diminished the need to reposition. As the war proceeded and medical need and sophistication increased, canvas huts became wooden buildings complete with bacteriological laboratories and dedicated radiology facilities. The sick and wounded were transferred to large General Hospitals located further from the fighting if necessary. From the General Hospitals soldiers were sent to back to base depot for reassignment, a convalescent camp to further recovery before some sort of reactivation, or home.

At each level of care, prompt evaluation and disposition were of utmost importance to ensure adequate space for the next wave of wounded soldiers. Convoys at the General Hospital arrived carrying between 50 to 500 patients, with up to 800 admissions in a day. Shattuck distilled the purpose of this wartime military health system: "Two military aims were never lost sight of: Firstly, to have at all times enough vacant beds to accommodate a convoy; secondly, to hold in France and return to duty every man who would be fit for it within the time during which he could be kept in hospital or a convalescent camp."[18]   This need for promptly vacating beds generated potential problems in establishing ongoing care of fractures to avoid preventable deformities from immobilization or suboptimal fixation.

Back in England, a large part of the orthopaedic work involved rehabilitation, and it was to this task that the first Americans arriving were assigned. This effort included curative workshops, where rehabilitating patients were taught new and productive skills. Activities included carpentry, leatherwork, cobbling, and tailoring. Some made and mended deep-sea fishing nets, while other produced splints and shoes, raised new buildings, and (in London) served in a patients' orchestra that played for nearby neighborhoods.[19] These had a multifaceted purpose: a means to pass the "monotonous period of healing"; a form of physical therapy, and a convenient way to cheaply produce splints, braces, repaired clothing and furniture for the hospital.[20] Orthopaedic surgeons oversaw this therapy alongside their mechanical and surgical treatments.

When Goldthwait accompanied the first group of orthopaedists to Europe, his purpose was to gather information and return home to aid in the formulation of orthopaedic policy as American troops prepared for deployment. The greatest challenge appeared to be

---

[18] Ibid 251-252.

[19] Osgood, The orthopedic centers of Great Britain and their American medical officers.

[20] Goldthwait, The place of orthopedic surgery in the treatment of war casualties, 453.

deficiencies in the initial care at the front. Goldthwait commented, "Sir Alfred Keogh, the Director General of Medical Services (equivalent to our Surgeon-General of the Army), urged me in our planning to start our reconstruction work in the front-line trenches, thus saving the unnecessary deformities which the orthopaedic surgeon was forced to correct later."[21] The immediate need of transfer, combined with the sheer numbers of patients, created challenges for the treatment of fractures.

Moving towards the front: beginnings of involvement with acute care

Fracture care was not an immediate reward for orthopaedic surgeons who served in Europe. Who would care for fractures is a struggle that began in earnest during the Great War, but far outlasted it. The first entry of American orthopaedists into front-line care came through establishing standardized splints and training for their application. Osgood and Allison were among those who finished a Splint Manual in October 1917, which was to be distributed to every medical officer either before leaving the United States, or upon arrival in Europe. Once back in Europe, Goldthwait traveled to high-level meetings while distributing the manuals. American soldiers were already arriving in France, though they would not come in large numbers until 1918. Experienced orthopaedic officers began being transferred from England to new US Base Hospitals in France, and were replaced by others from the United States. In this way new arrivals were supervised and acclimated with experienced surgeons in England before moving closer to the front.

For the splint manual, principles of traction and fixation were joined with simplicity of construction, transport, and ease of training. By establishing standard types, the manual accomplished a great reduction in the variability, number, and complexity of apparatus:

---

[21] Goldthwait, *The division of orthopaedic surgery in the AEF*, 14.

The Board believes that with the three types of wire-ring traction and counter-pressure fixation splints embodying the Thomas principle, the Jones "Cock-up," "Crab" wrist splint, the long interrupted Liston splint with adjustable foot piece, an anterior thigh and leg splint, Hodgen type, the Cabot posterior wire splint, the wire-ladder splint material, light splint wood, and plaster of Paris bandages and Bradford frames, treatment of all bone and joint battle casualties may be efficiently carried out at the Front, and if necessary in base hospitals.[22]

The orthopaedic leadership inserted themselves closer to initial injury care by dictating standard splints and equipment lists for combat divisions. One orthopaedic surgeon was placed in each division, to give each medical department staff instructions in the use of splints and the types of dressings to be used, as well as advice to the fighting men on "... proper use of the body, so that they would be able to meet the requirements of combat training."[23]

Goldthwait describes the extent of responsibilities of an orthopaedic surgeon in a combat division:

It was the intent, in each of the Combat Divisions, that the orthopaedic surgeon, as a member of the Staff of the Division Surgeon, would assist the regular medical officers of the Division with the special orthopaedic cases, and also see that the splints and types of dressings which had been decided upon were understood. It was also expected that the stretcher bearers would be instructed in the application of the splints. Still further, since a large number of men were in very poor physical condition, the body posture being very poor, a standardized talk was prepared which is was expected would be delivered by the orthopaedic surgeon to the groups of men in the Division Area

---

[22] Ibid 35.

[23] Ibid 46.

wherever it was possible for the orthopaedic surgeon to find them.[24]

The prepared talk included advice on footwear, posture, and lifting with the back straight. While orthopaedists were involved in clinical work, they also took on responsibilities in preventive orthopaedics and teaching splinting techniques. Another Harvard orthopaedist, Z.B. Adams, took command of a temporary convalescent camp for treatment of foot and back problems in arriving soldiers. Treating amputees was also delegated in part to orthopaedic surgeons, and Dr. Philip D. Wilson was sent to specifically study the problem of artificial limbs. Several niches for orthopaedic involvement developed, but acute fracture care was not initially among them.

As more American soldiers arrived in France in early 1918 and base hospitals began receiving more wounded, there were tensions between orthopaedics and general surgery. The problem discussed earlier, of imperfect initial treatment and rapid transfer of patients among various clinicians resulting in preventable deformities, led the orthopaedic leadership to push for standardized treatment under their supervision.[25] In discussions between Goldthwait and the Director of General Surgery, there was eventual recognition that most general surgeons had little experience in the long-term care of joint and other fracture cases. A letter to the Chief Surgeon, written in consultation with Goldthwait, was nearly replicated in an order that momentously shifted the provision of care for musculoskeletal injuries. With the approval of General Pershing, commander of the American Expeditionary Forces, Circular 11 is an underappreciated landmark in the history of orthopaedic surgery. Dated March 8, 1918, its clarity and impact justify its full inclusion:

---

[24] Ibid 53.

[25] Ibid 31.

The following instructions are issued for the guidance of all Medical Officers: -

1. Injuries to the bones and joints, as well as of the muscles and tendons adjacent to these structures, represent a large percentage of the casualties of both the Training and the Combat Periods of an Army.

2. To restore useful function to these injured structures is one of the purposes of the Medical Organization of the Army. The problems involved in this have to do not only with the cleansing and healing of the wounds, but also with the restoration of motion in the joint or strength to the part. This latter part naturally follows the first, but it is essential that the first part be carried out with reference to that which is to follow. Unless this second part of the treatment, the restoration of strength and motion, is carried out, much of the first part is purposeless.

3. To insure to the man not only the proper treatment for this type of injury, but the proper supervision until he is as fully restored as possible, necessitates some form of radial control that makes it impossible for a man to be overlooked in inevitable transfers, from service to service, or hospital to hospital.

4. Since so much of the ultimate result in these conditions depends upon orthopaedic measures after the first treatment of the wounds has been carried out, the following will govern: -

The Director of Orthopaedic Surgery is responsible for the treatment of the injuries or diseases of the bones or joints, exclusive of the head and face.

He will be held responsible for the treatment of injuries or diseases of the ligaments, tendons or muscles, that are involved in the joint function, of the extremities.

Officers attached to other Divisions may operate upon and treat such conditions, but the Division of Orthopaedic Surgery, through its Director, will be held responsible for the character of the treatment and for the final results.

It is expected that the direction and supervision of the treatment here indicated will be carried out, in so far as is possible, in co-operation with the Director of the Division of General Surgery.

5. To carry out the instructions of this circular, the Director of the Division of Orthopaedic Surgery will arrange so that representatives of his Division will see all cases of the nature described, to determine whether or not their management is proceeding satisfactorily so as to obtain the best possible results. These representatives will report to the Commanding Officers of the hospitals in which such patients are being treated and their services as consultants will be freely utilized: any recommendation made by them as to change of treatment, transfer to some other professional service, or hospital will ordinarily, if the military situation permits, receive favorable consideration.

6. It is not the intention of this order to interfere with the routine work of hospitals, but to insure to the soldier proper supervision during the time of his treatment and the period of his convalescence.[26]

With a commander's signature, the specialty dramatically enlarged its authority and boundary. Of note, however, the circular specifically did not preclude general surgeons from operating on these cases. Orthopaedics was given an increased role of supervision and essentially rehabilitation of musculoskeletal injuries, but not a clear authority over all initial operative management.

Provision of injury care under orthopaedic oversight

One of hospitals receiving wounded from the front was U.S. Army Base Hospital No. 6, located in Bordeaux, France. It had an orthopedic service, led by Dr. Henry C. Marble of MGH, who later wrote its history. He describes how the new regulations manifested at the base hospital:

[26] Ibid 75.

> *As the summer of 1918 wore on, these wards became filled with bone and joint cases, and it was at this time that orders were issued designating them as the Orthopedic Department, with the writer as head of the service. All the appliances which had been prepared by Captain Adams were now put in use, and here in this small group of wards could be found between three and four hundred cases of bone injury. ... In some wards there were as many as fourteen or fifteen compound fractures of the femur, all in Balkan frames with weights, pulleys, and splints, and with septic wounds to dress.*[27]

There were hundreds of patients, with all varieties of fractures, and all newly under the control of the Orthopedic Service.

So-called Reconstruction Aides were assigned to the service. One group of these allied services conducted the massage and physiotherapy, and the other were occupational therapists who organized the vocational training and curative workshops. "Some of the wards were literally bee-hives of industry, and the wounded men who had been lying in bed, some of them for months without occupation, or any way of filling in the long hours, now had work to do. They made toys, baskets, and paintings. Distinctly the whole tone of the ward was changed, and these idle hours became most profitable."[28] This collaboration with physical and occupational therapists, working under the direction of orthopedists, will later become a point of contention with the as-yet underdeveloped field of physiatry. But on Base Hospital No. 6, following the circular, orthopedists had clear control over all aspects of rehabilitation.

---

[27] Marble HC. Orthopedic surgery, in: *The History of U.S. Army Base Hospital No. 6 and its part in the American Expeditionary Forces, 1917-1918.* (Boston: Thomas Todd Company, 1924) 91.

[28] Ibid 94.

The following account is illustrative of soldiers' experience towards the end of the war, both of the treatment received and the degree of organization. The full quote gives a sense of the patient progression, and the course of their care in the hands of orthopedists.

> *Patients usually arrived direct from a field hospital or from a hospital in the advanced field zone. They came on hospital trains, and were transported from the trains in ambulances to the hospital receiving room. Here they were sorted by the Receiving Officer, and the bone and joint injuries were assigned to the Orthopedic Service. Usually when the patients arrived in the wards all dressings were taken down, and records of the case, which usually came in a small envelope around the patient's neck, read and studied. The dressings were reapplied, and the patients were sent at once to the x-ray department for pictures. The splints were readjusted and the patients generally suspended in Balkan frames. The usual routine was to treat the open wounds by the Carrell Dakin method, and the subsequent dressings were as a rule carried on by the nurse in charge of the ward. The splinting and the care of the Balkan frames and the other apparatus on the patient was detailed to the men of the hospital corps. In many of these cases major surgical operations had to be done, cleaning up infected areas, removing shell fragments, and loose pieces of bone. Following these operations the after care consisted in reestablishment of the Carrell-Dakin routine and the resplinting of the members. The wounds being healed, and the alignment of the fractures being satisfactory, the patients were then loaded into litters and sent directly to hospital ships in Bordeaux Harbor for their transportation to the United States.*[29]

After operations, splints and dressings were re-applied by a Splint Team, which included a junior orthopaedic surgeon. Given the need for beds, transfers to large Hospital Areas in the Rear were done as soon as possible. "Most of them, even though seriously wounded,

---

[29] Ibid 95.

were moved within three or four days, the less seriously wounded within a day or two."[30]   There the assessment was made as to disposition: return to combat duty, assignment to non-combat duty, or a return to America for further rehabilitation.   Hospital trains (equipped even with a simple operating room) facilitated this enormous undertaking of movement; more than 300,000 injured soldiers were transferred by train between July and November of 1918.[31]

Under the authority of the circular, this effort across countries, continents, and oceans was given orthopaedic oversight. Goldthwait proudly recalls the final result of his efforts:

> *With this type of organization, as soon as the patients reached the Evacuation Hospital where the first surgery was performed till the time the patients were ready to be put on the transports for sailing for home, they were under the constant supervision of trained orthopaedic officers, who reported in turn to either the Assistant Director of Orthopaedic Surgery or to me personally.*[32]

Orthopaedics:
The Principle of Restoring Locomotive Function

The preceding overview describes the course of events in World War I. From uncertain beginnings, orthopaedists eventually came to have oversight over all musculoskeletal injuries, including fractures. Understanding how the field articulated itself in the context of the war begins the analysis of how this transformation occurred.

---

[30] Goldthwait, *The division of orthopaedic surgery in the AEF*, 86.

[31] Ibid 87.

[32] Ibid 89.

From his first experiences in the war, Osgood formulated a set of essentials for efficient orthopaedic work in a war hospital. Chief among these was, "... a person interested in orthopaedic surgery, and in those mechanical principles which have to do with the fixation of inflamed joints and broken bones, and also with the restoration of function in stiffened joints and the obtaining of proper final alignment in fractures."[33] While the injuries, and their acuity, were previously outside of orthopaedics scope, an understanding of locomotive function could be applied:

> The type of injury to be treated had rarely ever been seen by the orthopaedist any more than by the general surgeon, but the basic principle of training made the orthopaedic surgeon see from the very beginning an end result, and the special case simply demanded the adaptation of well understood principles to the War casualty.[34]

A focus on the 'ultimate end result' of the injured part became the common denominator of orthopaedics.

This conception of orthopaedics became prominent during and immediately after World War I. During the war, Dr. Arthur Keith of the Royal College of Surgeons delivered a series of lectures subsequently published as *Menders of the Maimed: the anatomical & physiological principles underlying the treatment of injuries to muscles, nerves, bones, & joints*. Generated in direct response to the needs of treating the war wounded, its aim was to describe a "re-statement of the principles which underlie the art of Orthopaedic Surgery" by exploring its history.[35] It is now a sourcebook on the field's history and a foundational text encapsulating that period's understanding of the specialty. Keith defined the purpose of orthopaedics: "To effect

---

[33] Osgood, Orthopaedic work in a war hospital, 110.

[34] Goldthwait JE. The backgrounds and foregrounds of orthopaedics. *J Bone Joint Surg Am* 1933; 15:282.

[35] Keith A. *Menders of the maimed: the anatomical & physiological principles underlying the treatment of injuries to muscles, nerves, bones, & joints*. (London: Oxford University Press, 1919) vii.

the repair of the mechanical framework of the human body by all operations and appliances with that aim in view."[36] This description far displaces Andry's vision, as well as those by American orthopaedists in the late 1800s.[37] In response to the needs and opportunities of the war, a small but growing specialty recast itself anew.

An Early Appropriation of End Results in Orthopaedics, as applied to Fractures

The focus on 'end results,' and this particular terminology, suggests that Osgood and Goldthwait were drawing on the work of another MGH colleague, Dr. Ernest A. Codman (1869-1940). With the modern emphasis on outcomes studies and comparative effectiveness, Codman's early advocacy for following-up the results of treatment has been the well documented.[38] Although trained as general surgeon, Codman made notable contributions to orthopaedic interests including the radiographic assessment of fractures, anatomy and surgery of the shoulder, and for starting a bone tumor registry. After graduating HMS in 1895, Codman joined the surgical staff at MGH. From 1910 onwards he devoted himself to his 'end results system,' which refer to follow-up studies, the lack of which he found deplorable.[39] Codman's advocacy for assessing the outcomes of treatments led him to start his own hospital, aid substantially to hospital reform efforts, and infamously attack the leadership of the

---

[36] As quoted in: Osgood RB. *The evolution of orthopaedic surgery.* (St. Louis: The C.V. Mosby Company, 1925), 8.

[37] The debate over the definition of orthopaedics in this period is discussed in "Setting the Stage."

[38] Donabedian A. The End Results of Health Care: Ernest Codman's Contribution to Quality Assessment and Beyond. *Milbank Quarterly*, 67(2):234-235, 1989; Mallon B. *Ernest Amory Codman: the end result of a life in medicine.* Philadelphia: Saunders, 2000.

[39] The End Result System is fully described by 1918, in his book outlining results from a hospital he founded. Codman EA. *A Study in Hospital Efficiency: As Demonstrated by the Case Report of the First Five Years of a Private Hospital.* (Boston: Thomas Todd Printers, 1918). His interest predated this by at least eight years, as discussed in Mallon, *Ernest Amory Codman*, 51-52.

MGH in 1915. Assessments of Codman's End Results Idea, including by himself, have suggested that regular study of the final outcome of treatments was not readily initiated.[40] The successful adoption of end results as part of the principle of orthopaedics during World War I, however, illustrates early impact of his work.

Some of the lack of function and stiffness seen in soldiers at English hospitals early in the war seemed out of proportion to the initial wound. "Men who had received good surgical care at the Front," Goldthwait wrote, "were badly crippled when they reached England, much more than was justified by the original injury or by the first surgery. The aftercare, or the planning of the original care with reference to the ultimate well-being of the individual, became a matter of greatest importance."[41] The problem of injury care was framed as an inattention to the functional end result.

The rationale for orthopaedic involvement in fracture care did not depend on a particular surgical expertise, or an anatomical claim to authority over the extremities. Instead, Osgood discusses fractures in the context of preventing deformity:

> *A foot allowed to become fixed in equinus, a knee in extreme flexion, a hip in adduction, a wrist dropped, a straight elbow, a shoulder which is not abducted is a severe handicap and an entirely preventable one.*[42]

When ankylosis was likely, in the case of open and usually septic joint fractures, the immobilization position was all the more important to facilitate daily activities. The objective was to ensure that unavoidable ankylosis proceeded in a position of function, like

---

[40] The preface to his book on the shoulder suggests it would take generations yet for the end result idea to be accepted. Codman EA. *The Shoulder: Rupture of the Supraspinatus Tendon and Other Lesions In or About the Subacromial Bursa.* (Boston: Thomas Todd Co., 1934).

[41] Goldthwait, *The division of orthopaedic surgery in the AEF*, 7.

[42] Osgood, The orthopedic phases of military surgery, 99.

allowing the fingers to reach the mouth despite an immobile elbow joint. If careful attention is not paid to the position of immobilization, and if a commitment to maximizing future function not made, then serious and unnecessary disability resulted. Osgood believed the orthopedist would contribute to fracture care by always being concerned with future function.

The Case for Orthopaedic Involvement: Fitting Military and Economic Needs

Articulating the field as the experts on restoring function positioned orthopaedists to broaden their reach in the division of labor, including the treatment of fractures. Osgood had argued that orthopaedists had work to do in the war, and an opportunity to serve. In speaking before the College of Physicians of Philadelphia, he recalled,

> It was impressive to find at the American Ambulance in a general surgical service taken over by a new group of surgeons interested in orthopedic surgery that fully 25 per cent. of the cases needed some definite preventive or corrective measure to ensure an end-result entailing the least possible crippling.[43]

The added value of his new team, in direct comparison to the general surgeons, was in the recognition, prevention, and treatment of functional disabilities in the soldiers.

In pressing his case for involvement in the war, he spoke to the Section on Orthopedic Surgery at the June 1916 Annual Session of the American Medical Association. He stated his case directly:

> Our thesis is to be that orthopedic surgery has a very large part to play in [1] assuring physical efficiency in the ranks; [2] in

---

[43] Osgood, The orthopedic phases of military surgery, 98.

conserving and restoring the function of the locomotive apparatus of the wounded; [3] in providing the physical possibility and perhaps reorganizing the means by which the war cripples may become happy, productive, wage earning citizens, instead of boastful, consuming, idle derelicts.[44]

From evaluating footwear and posture at the basic training camp, to limiting stiffness and keeping a focus on ultimate functional outcome at the base hospital, to directing the occupational therapy of soldiers returning with injuries, Osgood pushed for a pervasive involvement of orthopedics in the care of the armed forces. The potential benefit to the army and government coffers was far from overlooked.

Courting the approval and cooperation of the Federal government, by positioning orthopaedists as a solution to the problem of wounded soldiers, plays prominently in their accomplishments. In Osgood's arguments for orthopaedic involvement, he made a case based on serving the interests of the patient as well as the army and nation. This sentiment began to be expressed forcefully by the orthopedic community in the run-up to American involvement in he war and the months afterward.

Goldthwait was particularly blunt in these assessments during the war:

Within a few months, it became evident that the mere saving of life unless it was followed by, or associated with, attempts to restore function to the damaged part, represented a burden upon the nation that was so tremendous, and an economic waste that was so great, that the very existence of the nation was jeopardized... In fact it was equally obvious that unless something could be done, it was better, both from the point of view of the man and of the nation, to let him die upon the battlefield.[45]

---

[44] Osgood RB. Orthopedic surgery in war time. *JAMA* 1916; 67:418-20.

By all accounts, the Federal government was quickly supportive of the orthopaedic organizations in efforts to prepare for war.[46] Even before the United States entered the conflict, there were already 600,000 wounded British soldiers.[47] Estimates of the percentage of musculoskeletal injuries requiring treatment ranged from 30-50%.[48] There was widespread concern in governments of Europe that the wounded represented "a center of unrest," and "... that unless something could be done to improve their condition, or at least to have them feel that the Government had done its best for them, these individuals would become centers of revolution, and that no Empire or Nation could survive that."[49] The rapid deployment of orthopaedists was one of several early actions illustrating Federal government and military support of the field's potential contribution.[50]

During the war, the industrial and economic benefits to patient and nation from restoring and conserving function were explicit, as seen in the following report on a hospital's results:

> *Of 1,350 patients discharged in a given period, about a thousand were sent back into the army as Class A, Class B, or Class C men, and of this number 294 were Class A men. The remaining 350 cases were sent back into civil life better and*

---

[45] Goldthwait JE. The place of orthopedic surgery in the treatment of war casualties, 450.

[46] This support is mentioned in several articles and histories of the period. A characteristic description of the Federal willingness to accept orthopaedics into the war effort is found in: Goldthwait JE. The place of orthopedic surgery in the treatment of war casualties.

[47] Osgood. The orthopedic centers of Great Britain and their American medical officers.

[48] Goldthwait JE. The place of orthopedic surgery in the treatment of war casualties.

[49] Goldthwait, *The division of orthopaedic surgery in the AEF*, 7.

[50] The existing presence of the AOA, and AMA Orthopaedics Section, were instrumental in attracting and directing governmental support. The Orthopaedic Advisory Board of the Army was made up "mainly of ex-presidents of the AOA, and of those representing the Orthopedic section of the AMA." Cooter R. *Surgery and Society in Peace and War: Orthopaedics and the Organization of Modern Medicine, 1880-1948.* (The Macmillian Press Ltd: London, 1993) 129.

industrially trained so that although crippled, they were still of use to themselves and the community. When it is realized that all of those cases would ordinarily have been discharged from the army as hopeless pensioners, the significance of the results is obvious and one is not surprised that the nation is demanding more and yet more of this kind of work.[51]

Likewise, at Shephard's Bush in London, there was pride in returning 1000 of the first 1300 patients to army service.

Increased orthopedic involvement served governmental interests in preventing a massive class of crippled young men on lifelong disability support. The military was responsive to the possibility of returning injured soldiers to active duty. By redefining the field in relation to the principle of maximizing function, orthopaedists gained oversight of the treatment and rehabilitation of injured patients into productive and self-reliant members of society, if not soldiers suitable for return to the front.

The Need for Systematic Care

During the course of the war, clinical experience led to the selection of particular antiseptic solutions, certain splints, and general treatment plans in military surgery. The fundamental advance of wartime care of musculoskeletal injuries, however, was organizational: case records hanging on the neck, standard radiographs, assurance of adequate alignment, and supervised management through the process of rehabilitation. The perceived necessity of such an organization arose from both the scale and the ever-changing location of care delivery, as outlined previously. Goldthwait was proud of the delivery system that emerged: "The perfection of the organization was remarkable; the wounded were being put on the port side of the boat, while the troops which had

[51] Goldthwait, The place of orthopedic surgery in the treatment of war casualties, 455.

just come over were being unloaded on the starboard side."[52] While a laudable logistic feat, it must have been a foreboding scene to the freshly arriving troops. Orthopaedists argued for and gained control of a coordinated system, centered on a sustained concern with deformity prevention and functional maximization.

The logistical complexities of the war were used as justification for a division of labor. General surgeons or other groups could have provided the sort of continuous care that was advocated, but the orthopaedic leadership articulated their field as possessing the functional focus to maximize outcomes for soldiers and society. Based on this experience, orthopaedists after the war would argue for continued specialized treatment of fractures. Debate on the merits of such a system, and who should control it, would come to the foreground in the interwar period.

Training the Principle: Orthopaedic Imperialism

Advocating for the functional rehabilitation of soldiers by orthopaedic surgeons was not by itself responsible for the field's success. A purposeful increase in practitioners also played a pivotal role. As indicated by England's initial request for orthopaedists, there was a significant need for surgeons able to treat and prevent deformities. It was consistently mentioned that orthopaedic work on a massive scale was necessary:

> One of the most outstanding features was the universal feeling that the work, in Great Britain, was largely a problem for orthopaedic surgery. Different estimates were given as to the percentage of cases that were requiring this type of care, but all placed the percentage very high.[53]

---

[52] Goldthwait, *The division of orthopaedic surgery in the AEF*, 92.

[53] Ibid 13.

Upon their arrival, the lack of personnel meant that their help was all the more appreciated, and the opportunities all the more forthcoming. When Goldthwait first returned to Washington, DC to report to the Surgeon General, of immediate priority was the training of more men in this functional principle who could help treat the substantial number of soldiers. Goldthwait concisely articulated the key principle, and argued that orthopaedists are the answer: "For such work, naturally, men who we trained to think of results in terms of function were needed, and it was here that the orthopedist found his opportunity."[54]

To fully capitalize on this dearth of surgeons, a training program was needed. Back in the United States steps were taken to address the gap in human capital. In July of 1917 the AOA officially offered the services of its membership to the Surgeon General, but its size was not sufficient to meet the demand.[55] Dr. Robert Lovett, then the John B. and Buckminster Brown Professor of Orthopaedic Surgery at HMS, organized and taught part of a nationwide effort to train additional surgeons the basics of orthopaedics. In an article from the *Journal of the American Medical Association* in 1918, Lovett relates his observations from five intensive courses designed to train military surgeons. [56] Three were taught at the Harvard Graduate School of Medicine, with two given at the Army's medical school in Washington, DC. The curriculum was standardized and expanded to courses in Philadelphia, Chicago, Oklahoma and Los Angeles. A focus on principles was Lovett's dominating theme. By the end of the course Lovett wanted each student to be capable of stating his treatment objectives in mechanical terms, to understand the principles involved, and be familiar with accepted forms of apparatus: "[Lovett] had the feeling each time this one department was completed that the student had been furnished with an alphabet by which to spell out the conditions as he met them rather than an incomplete encyclopedia, the items of which he might or might not

---

[54] Goldthwait,The place of orthopedic surgery in the treatment of war casualties, 451.

[55] Goldthwait, *The division of orthopaedic surgery in the AEF*, 8.

[56] Lovett RW. A plea for more fundamental method in medical teaching. *JAMA* 1918; 70:1070-2.

remember correctly."[57] Over 700 officers completed courses at the various sites.[58]

This rapid influx of orthopaedists, trained and subsequently experienced in treating war injuries, was crucial to the field's growth. The war was seen as a chance for expansion to entirely new areas, rather than simply an increased load of established work. Osgood stressed this unique chance to fundamentally broaden the boundaries of orthopaedics:

> We believe orthopedic surgery in times of peace is a most comprehensive specialty; in war time the possibilities of its helpfulness are still greater. The fact that these possibilities are not always recognized should make orthopedic surgeons seek opportunities to demonstrate them. If this opportunity is accepted it may well mark an epoch in the history of the specialty.[59]

Accomplishing this required a willingness of the profession to open its doors to new members. Far from putting up substantial barriers to the field, the courses by Lovett and others showed an eagerness to grow in numbers. This expansionist mindset was supported by AOA Presidents in the years immediately before the war, and was only taken further during the conflict.

In contrast, Robert Jones would come to regret the nearly complete lack of British students who trained in orthopaedics during the war. Both Jones and other commentators suggested that the relatively small numbers of practicing British orthopaedists in the interwar period dampened the possibility for gains in position and scope of

---

[57] Ibid

[58] US Official Medical History. Volume 9: Surgery, 552.

[59] Osgood, Orthopedic surgery in war time, 420.

practice.⁶⁰ We will see in the next chapter that when the war ends, freshly trained American orthopaedists were unwilling to give up the field, and they spread around the country with their accumulated experience in both acute and chronic treatment of adult musculoskeletal conditions. The ability and willingness to quickly train and send new members of the profession to Europe capitalized on a unique window of opportunity to gain experience treating the multitude of injured soldiers under the banner of orthopaedics.

Conclusions, and Looking Ahead

In articulating the argument for broader powers for orthopaedic surgeons, with control over restoring function from the time of injury through all the steps of surgical treatment and rehabilitation, orthopaedics underwent a transformation of its meaning. Initial doubts as to its usefulness in a war hospital were replaced with excitement at the opportunity to contribute. To succeed, the field shifted from dealing with chronic cases and a predominant pediatric base to a principle of restoring function to any pathology of the 'locomotive apparatus.' Leaders like Osgood and Goldthwait proposed that their field was the answer to the problem of massive numbers of wounded soldiers. In doing so, they garnered the support of the government and military by suggesting their efforts could return injured soldiers either to the battlefield, or else to productive employment free of lifetime dependence.

Once allowed into the wartime system of medical care, orthopaedists pressed their case further, suggesting that continual supervision and treatment from the initial injury all the way throughout rehabilitation could maximize the ultimate function. By the end of the war, orthopaedic surgery had received official management of all injuries to bones, joints, muscles and tendons. For the first time, fractures were managed by orthopaedists on a wide scale, though general

---

⁶⁰ Cooter, *Surgery and Society in Peace and War*, 130-1.

surgeons were still doing many of the operations. This represented a wide-ranging expansion of the field's scope. A simultaneous eagerness of the profession to expand its ranks facilitated the generation of critical human capital to do the work, and generated experience in treating these injuries. The Surgeon General circular was the culmination of efforts by the field of orthopaedics to recast itself as masters of restoring function, with an influx of members, who could aid the military and government by minimizing patient-soldiers' disability and maximizing their return to combat or industry.

The war represented a window of opportunity that the field of orthopaedic surgery utilized to undergo another transformation in identity. But with peace came the necessity of transitioning wartime efforts to civilian practice. In concluding his history of the Division of Orthopaedics, Goldthwait argued for orthopedic need beyond the theater of war. He suggested, "... no one can see the work and realize that the existence of it is to insure to the men, not the mere saving of life or the mere healing of wounds, but the restoration of useful function in the damaged part, without being conscious that the same needs exist today in the injuries of civil life."[61] He believed that the training and experience gained by orthopedic surgeons in England during the war was of "inestimable value" to treat American wounded soldiers, and be prepared upon their return home to care for anyone crippled by disease or industrial injuries in peacetime.[62]

The potential role in treating industrial mishaps, the prospect of which predated the war, entered center stage upon the cessation of hostilities. Navigating conflicting claims to this growing medical need, and managing a suddenly expanded field of orthopaedics, would again force a re-evaluation of the scope and grounding of orthopaedics.

---

[61] Goldthwait, The place of orthopedic surgery in the treatment of war casualties, 455.

[62] Ibid 456.

From War to War:
Becoming Surgeons of the Extremities and Spine

Introduction

World War I brought orthopaedic surgery into a new prominence, and enlarged its boundaries to include the acute care of musculoskeletal injuries and their rehabilitation. How to adapt these gains to civil life became a challenge for the field and its influx of new practitioners. The possibility of controlling the care of industrial injuries became a goal immediately after the war. This was a growing area, made profitable by salaried positions in companies as well as workers' compensation laws. Meanwhile, advances in public health decreased the prevalence of former bedrocks of orthopaedic practice such as skeletal tuberculosis and rickets. However, treating the 'soldiers of industry' meant competition with general surgeons, as well as physicians who were trying to specialize as industrial surgeons.

The experience of the MGH Fracture Clinic illustrates that while orthopaedists gained some control of fracture care, they had to settle for shared responsibility with general surgeons. Extending their principle of restoring function could no longer define the field, as increased attention to end results became more commonplace. To demarcate their work, a push was made to define orthopaedics anatomically and surgically, as the field that operates on the spine and extremities. In doing so, the rhetoric in this period shifted to accept a role as a branch of general surgery. Resolution as to who would control fracture care came during World War II, but the gains made for orthopaedics in operative treatment emerged alongside a gradual loss of dominant control over rehabilitation.

A Necessary Readjustment

The AOA Presidential address in 1919 was appropriately entitled "Readjustment to changing conditions."[1] The speed with which orthopaedics gained in stature and scope is stressed:

> *The war has done more to bring Orthopedic Surgery into its true inheritance than would have been accomplished by other agencies in many years; if indeed anything else would ever have brought about such complete emancipation of our specialty.*[2]

While the war provided a unique window of opportunity, how to move the field forward in the sudden peace was cause for anxiety. "Our specialty is," the AOA President continued, "in a period of transition, and the difficulties and confusion which are more or less inseparable from all processes of readjustment to changing conditions cannot be wholly escaped..."[3] There was now a cadre of new orthopaedic surgeons, trained in military surgery. They were not likely to simply put aside the experience gained in Europe.[4]

But in the absence of war, what would busy these orthopaedists? As was appreciated, many of the quickly trained officers did not have

---

[1] Galloway HPH. Readjustment to changing conditions. *Journal of Orthopaedic Surgery* 1919; 1(7):395-401.

[2] Ibid 396.

[3] Ibid 396.

[4] Galloway stressed the connection between the war experience of new orthopaedists and their continuation in civil life: "The sudden enormous increase of orthopedic work created by the war, together with the expansion of the former boundaries of this specialty, created a demand for trained orthopedic surgeons which could not be met from the ranks of recognized specialists; consequently many young general surgeons have, during the war, enjoyed exceptional opportunities for acquiring a familiarity with military orthopedic surgery, and it is almost inevitable that not a few of these, enthralled by the intensely interesting work of physical reconstruction will desire to be regarded as orthopedic specialists on their return to civil practice." Ibid 397.

experience in clubfoot, skeletal tuberculosis, scoliosis or poliomyelitis.[5] Trauma to the extremities and spine were more familiar to them than mechanical apparatus for congenital deformities.

The Promise of Industrial Surgery

There was an immediate consensus that the path forward for orthopaedics would be intimately related to the field of industrial surgery. As Osgood commented soon after the war's end, "There is a field in which are produced crops of injuries which differ only in degree and not in kind, from war injuries... The field is Industrial Surgery."[6] The rising tide of wounds in factories and railroad construction, as well as from automobile collisions, provided a chance to apply the same principles and treatments that this generation had learned in Europe.

Physician interest in treating industrial injuries preceded the Great War by decades. Railroad companies, steel makers and other manufacturers had an interest in preventing and treating injuries of their workers. Their reasons were similar to why the government and military wished to return wounded soldiers to active work or at least minimize permanent and costly disability. The Interstate Commerce Commission found in 1890 that one in every twenty-eight employees was injured on the job. Surgeons had been employed by railroads increasingly after the Civil War, and as their construction

---

[5] This knowledge gap in the conditions of childhood was recognized: "many a surgeon who could do the most difficult bone-grafting operation with consummate skill might find himself floundering discreditably in his attempt to treat cases of scoliosis, infantile club-foot, the disabilities resulting from infantile paralysis, and many of the tedious tubercular bone and joint conditions which demand the illimitable patience and mechanical resourcefulness of the older school of orthopedic surgeons." Ibid 398.

[6] Osgood RB. The orthopaedic outlook. *Journal of Orthopaedic Surgery* 1919; 1(1):5. This article, being the first piece in the first issue of the renamed journal, is a landmark piece on orthopaedic thought after the war.

moved westward into rural areas, doctors were retained or clinics established to provide care for their employees. [7]

There were several reasons why orthopaedists were eager to begin treating wounded industrial workers. If willing to move to a rural area, an orthopaedic surgeon back from Europe could receive a salary that came from deductions from workers' wages. Workers' compensation laws had been established in the states in the 1910s.[8] This meant that some working-class employees were able to pay for a surgeon's services; prior to these laws, itinerant single men were exactly the sort of socially isolated, poor patient for which charity hospitals like MGH were established. These workers could now be a source of income, rather than a necessity of charity.

Changing Epidemiology of Pathology

Opportunities in industrial surgery contrasted with declining work in traditional orthopaedic domains. Not only were a segment of the new generation of orthopaedic surgeons unfamiliar with treating the conditions of childhood that were the prior bedrock of practice, but some of these illnesses were also becoming less common. The landscape of illness was changing, and would continue to do so primarily as a result of gradual and remarkable advances in treating and preventing infectious diseases.

The first decline was seen for skeletal tuberculosis. At the Children's Hospital in Boston, which had built its reputation on treating Pott's disease, the change was rapid and dramatic: "A decline which had begun when the 136 admissions for spine and hip disease in 1921 fell

[7] The growth of industrial surgeons prior to World War I is discussed as it related to industrialization, transportation, and the beginnings of company-based insurance in: Starr P. *The Social Transformation of American Medicine.* (USA: Basic Books, 1982) 200-9.

[8] Guyton GP. A brief history of workers' compensation. *The Iowa Orthopaedic Journal* 1999; 10:106-110.

to 80 in 1924, and 67 in 1925, had reached bottom... In 1936 there was exactly one child with tuberculosis of the spine admitted.[9] Securing the milk supply from bovine tuberculosis, and general increases in public sanitation, are credited with this decline.

Another disappearing condition was rickets. It was routine from the 1890s through the 1910s for discussions of the etiology and surgical treatment of rickets and osteomalacia to be featured in the reports on progress in orthopaedics. A prominent orthopaedist would recall in 1948, "During our professional lifetime we have seen the disappearance of rickets, which in 1900 provided an enormous number of deformities for the orthopaedic surgeon to correct by means of Grattan's osteoclast."[10] Vitamin D and its function were discovered in the early 1920s, and as the efficacy of cod liver oil and sunlight to prevent the deformities of rickets largely eliminated this source of work for orthopaedic surgeons.[11]

Advances in public health, nutrition, and later vaccines[12] advanced a realization that treating the aftermath of infectious and metabolic diseases was not the future of orthopaedics. Attention was paid to chronic conditions, evidenced in Boston by the founding of the Robert B. Brigham Hospital in 1914 as a center for the study and treatment of arthritis. However, the revolution of effective arthroplasty would not be widespread until the 1960s. From the 1920s through the 1950s, orthopaedic treatment of trauma was

---

[9] Smith CA. *The Children's Hospital of Boston: built better than they knew.* (Boston: Little Brown and Company, 1983) 66.

[10] Harris RI. The future of orthopaedic surgery. *J Bone Joint Surg Am* 1948; 30:809.

[11] In the orthopaedic literature, a May 1923 report on orthopaedic progress mentions an article on a fat soluble vitamin that "would seem to be to regulate the metabolism of bone." Osgood RB, Soutter R, Low HC, Danforth MS, Bucholz HC, Brown LT, Smith-Petersen MN, Wilson PD. Twentieth report of progress in orthopedic surgery. *Archives of Surgery* 1923; 6:858-908. However, articles on cod liver oil as treatment and prevention of rickets predate this discovery.

[12] The aftercare of children with poliomyelitis would occupy the careers of many orthopaedists for decades. The breakthrough development of a vaccine in the late 1950s eventually eliminated this disease, and its significant musculoskeletal morbidity, in America.

increasingly seen as a necessity if the field was to continue its growth.[13]

An Ongoing Battle for Fracture Care

While industrial surgery was an understandable target for orthopaedic surgeons returning from Europe, they were not the only ones advocating to treat the 'soldiers of industry'. General surgeons, general practitioners,[14] and newly proclaimed specialists with their own journal were positioning for control of this growing field. This situation provides an example where a natural division of labor model does not fully explain the developments, as several systems of specialized treatment were possible and existed simultaneously.

In arguing for involvement in industrial surgery, the treatment of fractures was a critical element. Disputes over who should care for broken bones rekindled debate over specialization, as well as how to transition the lessons from the war. In their efforts to argue for orthopaedic involvement, the field would again re-articulate its boundaries.

The MGH Fracture Service

The provision of treatment for fractures at Massachusetts General Hospital exemplifies the developments of the interwar period. Initial

---

[13] The importance of trauma, and the falling rates of traditionally treated conditions, was recognized by the orthopaedic community. Cave EF. Trauma and the orthopaedic surgeon. *J Bone Joint Surg Am* 1961;43:582-9.

[14] A handbook entitled *The Treatment of Fractures in General Practice* was published in London in 1923. A review was published in the *Archives of Surgery*, though with skepticism on whether GPs have much experience in these treatments. Osgood RB, Allison N, Soutter R, Bucholz CH, Brown LT, Danforth MS, Low HC, Wilson PD, Smith-Petersen MN, Swaim LT. Twenty-third report of progress in orthopedic surgery. *Archives of Surgery* 1924; 8:918-60.

dominance by general surgeons gave way to shared control after WW1. While forming a productive collaboration, general and orthopaedic surgeons had divergent visions for facture care. Proposed systems of care included the treatment of fractures by generalists, a specialty of 'traumatism' including damage to bones as well as organs, and orthopaedic control of musculoskeletal injuries. Widespread public disputes, publications on outcomes, the adoption of routine radiographic assessment, and educational campaigns failed to generate a dominant authority.

An early name in fracture care at MGH was Dr. Charles L. Scudder (1860-1949). After graduating from HMS in 1888, Scudder completed a general surgery externship at MGH, eventually spending 56 years at the institution and holding appointments in surgery at HMS. A general surgeon by training, Scudder was allowed to focus on stomach surgery as part of attempts at the hospital to develop special expertise in certain cases. He published on the surgical management of ulcers, as well as topics in abdominal, urologic, and neurosurgery among others. As his career progressed, however, he became intimately involved in the treatment of fractures and dislocations.

Scudder's interest and eminence in fracture care as a general surgeon was representative of the pre-World War I period. His first book on fractures was published in 1900.[15] It was Scudder who founded the MGH Fracture Clinic in 1917.[16] It was only after the war that orthopaedists were able to levy their newfound experiences and prestige to gain a greater role in fracture care at MGH. A so-called 'Fracture Service' was established, which included orthopaedists. The MGH Annual Report noted, "Since July [1920] the fracture cases have had the benefit of closer cooperation with the Orthopedic

---

[15] Scudder CL. *The treatment of fractures.* (Philadelphia: W.B. Saunders, 1900). While Scudder published smaller volumes in 1898 (*The ambulatory treatment of fractures,* and a short monograph on comminuted fracture of each patella), the 1900 volume was 433 pages, with 585 illustrations. It went through more than ten editions, the last of which was published in 1938.

[16] Osgood RB, Soutter R, Low HC, Danforth MS, Brown LT, Wilson PD. Eighteenth report of progress in orthopedic surgery. *Archives of Surgery* 1922; 5:413-47.

Department. Dr. Osgood now makes a weekly visit with Dr. D. F. Jones, who has charge of fractures."[17] The history of MGH from that period states that Daniel F. Jones, a general surgeon, was the Fracture Service Chief, with Osgood as Associate Chief.[18] Under this arrangement, patients admitted for treatment with a fracture were rotated between the surgical and orthopaedic services.[19] The following year's annual report commented, "Cooperation between Dr. D. F. Jones and the Orthopaedic Department on the subject of fractures has proved a success."[20]

It has been argued that Osgood's own prestige and diplomatic personality were instrumental in this agreement to share responsibilities.[21] Yet even after Osgood's retirement the Fracture Service thrived. According to the MGH General Executive Committee's 1923 report:

> *Cooperation between the Surgical Service and the Orthopedic Service under its new chief, Dr. Nathaniel Allison, has been intimate and satisfactory. This applies particularly to the Fracture Service under Dr. D.F. Jones. ... Particular attention has been paid to fractured hip in the aged. With the newer methods of treatment, the resulting disabilities have been definitely lessened.*[22]

A testament to this collaboration was a conference on fractures, first held at MGH in 1922. In attendance were 27 physicians from across

[17] *One Hundred and Seventh Annual Report of the Trustees of the Massachusetts General Hospital, 1920.* (Cambridge, MA: The University Press, 1920), 118.

[18] Faxon N. *The Massachusetts General Hospital 1935-1955.* (Cambridge, MA: Harvard University Press, 1959).

[19] Brown T. *Orthopaedics at Harvard: the first 200 years.* (Boston: Massachusetts General Hospital, 1984) 16.

[20] *One Hundred and Eighth Annual Report of the Trustees of the Massachusetts General Hospital, 1921.* (Cambridge, MA: The University Press, 1921) 62.

[21] Brown, *Orthopaedics at Harvard: the first 200 years,* 16.

[22] *One Hundred and Tenth Annual Report of the Trustees of the Massachusetts General Hospital, 1923.* (Cambridge, MA: The University Press, 1923).

the country ("as far west as Pittsburgh and Chicago"), and a consensus syllabus by the general and orthopaedic surgeons on treating common fractures published in the *Archives of Surgery*.[23] These conferences, which grew in size and continued for many years, were only one component to a widespread educational campaign, coordinated with significant involvement from orthopaedic and general surgeons at MGH. A notable example of their prodigious output of publications was a voluminous 1938 book entitled *Experience in the management of fractures and dislocations, based on an analysis of 4390 cases, by the staff of the Fracture Service, Massachusetts General Hospital, Boston*.[24] Its creative table of contents consisted of drawings of the human skeleton, with the bones color-coded to a legend listing chapters on each group of fractures. The authors included prominent orthopaedic and general surgeons, many of whom served in World War I. Both the shared clinical duties, and the joint publications regarding developments in fracture care, attest to a growing but shared expertise.

Shared use of new technology and knowledge

By the interwar period, radiographs were in widespread use in American hospitals. This added ability in imaging fueled advances in the diagnosis and treatment of fractures, as well as the opportunity to analyze ultimate rates of bone union. A primer on fracture treatment prominently notes that failure to take x-rays could result in adverse jury decisions in malpractice cases.[25] However important, the use of radiographs and the knowledge gained from them was not isolated to orthopaedists. General practitioners and surgeons

---

[23] *One Hundred and Ninth Annual Report of the Trustees of the Massachusetts General Hospital, 1922.* (Cambridge, MA: The University Press, 1922) 132.

[24] Wilson PD, ed. *Experience in the management of fractures and dislocations, based on an analysis of 4390 cases, by the staff of the Fracture Service, Massachusetts General Hospital, Boston.* (Philadelphia: Lippincott, 1938).

[25] *Illustrated primer on fractures.* (Chicago: American Medical Association, 1931). In this book, booths from fracture conferences are recreated in pictures, with several signs cautioning the potential consequences of failing to take radiographs.

likewise made use of this technology, and could put the lessons learned to use in their clinical practice. The myriad educational campaigns served to disseminate these techniques. Unlike examples from cardiologists' early adoption of electrocardiograms[26] and ophthalmologists' initial expertise with the ophthalmoscope,[27] new developments in this period did not provide a single specialty the opportunity to claim unique authority over fracture care.[28]

Divergent Visions for Fracture Care

While the MGH Fracture Service was a joint venture, and congeniality expressed at least in official reports, the general and orthopaedic surgeons at that institution and elsewhere had vastly different conceptions of fracture treatment's ideal organization. Several distinct formulations of fracture care were proposed and defended, two of which will be discussed here. A prominent debate from 1925 illustrates themes from this period. That year, Sir Robert Jones gave a lecture entitled "How Can Fracture Treatment be Improved?"[29] From an orthopaedic perspective, he described the

---

[26] Fye WB. *American cardiology: the history of a specialty and its college.* (The Johns Hopkins University Press: Baltimore, 1996).

[27] Rosen G. *The specialization of medicine with particular reference to ophthalmology.* (New York: Arno Press, 1972).

[28] General surgeons would continue to figure prominently in technological and surgical advances in treating fractures. The revolution in techniques and theories of open reduction and internal fixation, which coalesced and spread via the Arbeitsgemeinschaft für Osteosynthesefragen (AO), was developed largely by general surgeons in Europe beginning in the 1950s. By this point, we will see, fracture care in the United States was securely in the domain of orthopaedic surgery. This further shows that special knowledge in new technologies was not the route to expertise in fracture care for American orthopaedists. This contrasts with the latter developments in arthroplasty, which were advanced by orthopaedists who then maintained control of these techniques - though recent changes in the law have allowed podiatrists in some localities to replace knee joints, which if spread would constitute a major encroachment of orthopaedic practice (see "Conclusions and Speculations"). While this monograph will not detail the technical developments in fracture care, a thorough account can be found in: Colton CL. Chapter 1: The history of fracture treatment. *In* Browner BD, Jupiter JB, Levine AM, Trafton PG, Kretteck C, eds. *Skeletal Trauma,* 4th edition. (Philadelphia: WB Saunders, 2008).

[29] Jones R. Lady Jones Lecture on crippling due to fractures: its prevalence and remedy. *BMJ*

inadequate fracture care in England and suggested reforms to improve treatment and education.

Drawing heavily from the war experience, he recalled the soldiers who returned early in the war with "...appalling deformities. Excessive shortening and malunion were the rule." Jones stressed that the greatest advance made in the war was not technological or surgical, but organizational.

> *Let me emphasize again that the success in the treatment of fractures during the later phases of the war was due to expert supervision, simplicity of apparatus, teamwork, segregation, and appropriate after-care. These are the lessons the war taught us. Are we going to apply them, and improve upon them, or are we to revert to the old bad way?[30]*

As part of this plea for organization, specialization remerged as a conflict in fracture care. Jones saw specialization as necessary:

> *There is sometimes a natural inclination on the part of the general surgeon to resist what he fears to be an encroachment of specialism. This, I am happy to think, is less evident than in days gone by. There is no more circumscribed a specialty than that of the surgery of the abdomen.[31]*

According to Jones, all fracture cases should be segregated, and under the control of a special surgical unit that dealt with inpatients as well as outpatients. Its director must be a surgeon with a career-spanning special interest and knowledge of fractures. He was not shy of his preferences regarding the choice of these directors.

> *If there exists an orthopaedic department, as should be the case in every teaching hospital, the chief of this department should*

1925; 1:909-13.

[30] Ibid 909.

[31] Ibid 913.

> *be one of the surgeons in charge of fractures... In an ideal scheme the fracture wards should form part of the orthopaedic service; this would simplify both the organization and the teaching.*[32]

This line of argument from Jones gives orthopaedists credit for the organizational successes of World War I, as well as their central principle of restoring function. From this perspective, if fracture care is to be improved in civil life such specialization must continue, and under orthopaedic control.

In America, this was well covered in the published series entitled "Reports of Progress in Orthopedic Surgery." The issues raised were not isolated to England, according to Osgood and the other Harvard surgeons who authored these reports. They wrote, "In England Jones' article set fire to tender that evidently had been long smoldering."[33] The summary of his lecture and the subsequent debate was so extensive, the editors felt obligated to explain, "We have reported the foregoing article and discussion at considerable length because both seem of great importance. Jones has called attention to a state of affairs that requires reform, and it exists not only in England but also in this country."[34]

Debate on this topic occupied a joint session of the surgical and orthopedic sections at the following British Medical Association annual meeting. Osgood was at the meeting representing American orthopaedics.[35] In response to Jones, some general surgeons argued against segregation of fracture care as an unnecessary specialization,

---

[32] Ibid 910.

[33] Osgood RB, Allison N, Wilson PD, Bucholz H, Soutter R, Low H, Danforth MS, Brown LT, Smith-Petersen MN. Twenty-eighth report of progress in orthopedic surgery. *Archives of Surgery* 1926, split between 12:151-168 (January) and 12:604-618 (February).

[34] Ibid.

[35] In joining the debate, Osgood described the experience of the Fracture Service at MGH as an example of a specialized service, with joint control by general surgeons and orthopaedists.

and took special issue to being designated as "abdominal surgeons."[36] Other general surgeons, while recognizing the benefits of specialized fracture wards, disputed Jones' position that orthopaedists should lead these units.

If fracture was to be specialized, that did not necessarily mean orthopaedic control. Scudder, as an MGH general surgeon with an interest in fractures and dislocations, led efforts to frame such treatment as needing specialized care at the hands of general surgeons. His vision of fracture care was expansive, and included fractures to the thorax with damage to the lungs or heart, as well as pelvic fractures with injury to the abdominal organs.[37] He noted, "This conception of the surgery of fractures, therefore, is seen to cover pretty completely traumatic surgery."[38] Continuing from this, Scudder argued, "A compete general surgical training and the ability to exercise that sound judgment which comes with experience is obviously necessary for the man who is to handle well the many difficult situations which may arise in any case."[39] By including damage to organs as well as bones, this view of fracture care is a specialty of traumatic surgery, and therefore is the domain of the general surgeon. "Fractures, traumatic surgery," he concluded, "should always remain a part of general surgery."[40]

While respecting the work done in World War I, Scudder presented a critique of orthopaedists: "The opposition, I take it, that obtains on the part of the general surgeon to orthopaedic men being assigned to the care of fractures rests solely upon the fact of their lack of general surgical experience."[41] Training quickly in orthopaedics was seen as a

---

[36] Gask GE. Discussion on the treatment of fractures: with special reference to its organization and teaching. *BMJ* 1925; 2:317-31.

[37] Scudder CL. Certain problems concerning fractures of bone. Annuals of Surgery 1921; 74(3): 2805.

[38] Ibid 280.

[39] Ibid 280.

[40] Ibid 285.

short cut for "yet untrained surgeons." Rather than orthopaedic control of fractures, Scudder had a different vision:

> *The time is coming when young men of this country will specialize in Traumatic Surgery. They will not be orthopaedic surgeons doing traumatic surgery. The will primarily be Surgeons of Traumatism.*[42]

Through leadership posts and educational campaigns, Scudder attempted to advance this position. The American College of Surgeons, founded in 1913, began a Committee on Fractures in 1922 with Scudder as its chairman.[43] Other committee members included orthopaedic surgeons such as Osgood and Allison (who had moved from Boston to St. Louis). This committee eventually established minimal standards for hospitals regarding equipment for treating fractures, required a responsible surgeon be designated for supervising the care of fractures, set principles for treating and teaching the treatment of fractures, and forced attention on the evaluation of end results. Annual meetings of the committee included clinical demonstrations of fracture and trauma care. Its *Illustrated Primer on Fractures*, first published in 1931 and enlarging in size and popularity through numerous subsequent editions, was central to the education of physicians.[44] To recognize Scudder's contribution, the ACS annual Oration on Trauma still carries his name.

---

[41] Ibid 285.

[42] Ibid 285.

[43] This group continued work done by a committee of the American Surgical Association that was started in 1911, following a presidential address on the treatment of fractures. As noted, pre-WW1 fracture care was dominated by general surgery, which had expressed interest in the subject since 1891. As interest in industrial surgery and the frequency of automobile collisions grew after the war, so did the work of the ACS. Estes WL Jr. The role of the American College of Surgeons in the Treatment of Fractures. *J Bone Joint Surg Am* 1956; 38:1165-7.

[44] *Illustrated primer on fractures.* (Chicago: American Medical Association, 1931).

While orthopaedists were involved in its early work, the location of the Fracture Committee within the ACS and Scudder as its chairman suggested that general surgery was the proper authority on fractures. The work of Scudder and the ACS in this period show the efforts by individual general surgeons and their national organizations to control the boundaries of fracture care from orthopaedic encroachment.[45]

These two mutually exclusive conceptions of fracture care argue against a natural division of labor. While disagreeing ultimately on the role of the orthopaedist, the visions advanced by Jones and Scudder agree that specialized treatment for fractures would improve care.[46] If the critique of orthopaedists was a lack of training in general surgery, then a path to their inclusion in fracture care was apparent.

Orthopaedics Reframed, Anatomically: Surgery of the Extremities and Spine:

The principle-based orthopaedics as articulated in the war ultimately facilitated greater oversight over musculoskeletal injuries. But that authority did not transfer in peacetime. The debates over specialized care of fractures, the large-scale education of general and orthopaedic surgeons in fracture treatment, and the experience at the MGH Fracture Clinic, reveal dual control over these injuries. Efforts to transform their field led to shifting the boundaries away from the

---

[45] A specialty of traumatism, including both fractures and injuries to the body cavities, has not materialized in the United States. As will be discussed below, an effort to form a Board of Trauma failed in 1957. Considering Scudder's opinion that fracture care essentially meant all of trauma surgery, it is appropriate that the Fracture Committee eventually became the still-active Committee on Trauma. However, as the 2009 Scudder Oration illustrates, it is given by a general surgeon on non-orthopaedic (meaning non-fracture) trauma and emergency care.

[46] This did not preclude a role for increased education of general practitioners given their necessary involvement in treating common injuries, especially in communities at a distance from major hospitals. This is commented on in: Scudder, Certain problems concerning fractures of bone.

principle of function and towards an anatomical, more surgical reconception.

Despite wartime distinction, the continued existence of orthopaedics as a specialty was called into question during this period. Immediately after the war Osgood was surprisingly lukewarm of the field's future: "The war has suddenly brought into prominence a small specialty. It is fair to say that many honest surgeons believe into too great prominence. Will it remain a specialty? Probably yes."[47] While still championing orthopaedics as "A specialty of function and not of anatomy," he continued by noting that position's weakness:

> *Perhaps it is too broad, too inclusive. Does not all medicine and surgery attempt to make straight! What right has a small body of men to claim it as their special task! The only right is because few general surgeons and few internists have considered it their task.[48]*

It begs the question: what happens once other physicians and surgeons incorporate a concern for function? As discussed above, the push by the ACS included a focus on after-care and investigation of end results, which negated some of the orthopaedic arguments for their unique role in treatment.

Osgood was not alone in questioning the ongoing existence of orthopaedics in this period. An AOA president cautioned immediately after the war that without change, "the line of demarcation between orthopedic and general surgery will in time become almost indistinguishable."[49] Goldthwait openly wondered whether orthopaedic practice was, "...shaping it for permanence or

---

[47] Osgood, The orthopaedic outlook, 4.

[48] Ibid 2.

[49] Galloway, Readjustment to changing conditions, 398.

for ending."[50] Despite its prominence in the war, there was concern as to whether the field would even continue as a specialty.

During this period of confusion over what their field would become, Osgood and others sought to respond to critics in general surgery while also bolstering a position compatible with orthopaedic control of fracture care.   In his 1921 AOA Presidential Address, Osgood began by recognizing the prior framework of orthopaedics:

> *We as orthopaedic surgeons will never escape from the necessity of training ourselves in the use of apparatus nor in developing the faculty of invention by which we may be enabled to devise physical means which shall most perfectly fulfill the requirements demanded by an understanding of bodily mechanics.*[51]

The Bradford frame, and the Thomas splint, are among the purview of an orthopaedist, yet the diction of 'escape' and 'necessity' here already suggest a pending separation or at least mitigation of such work. Then Osgood confronted the new reality directly:

> *But if we essay to be trusted with the knife, if we call ourselves surgeons, we must be so in very fact. We have often felt justified in criticizing the work of general surgeons in bone and joint lesions.   Our own work, therefore, must not be unfavorably compared with the best which General Surgery and the other surgical branches exhibit.*[52]

---

[50] Goldthwait, The backgrounds and foregrounds of orthopaedics, 279.

[51] Osgood RB. Backgrounds of orthopaedic surgery. *Journal of Orthopaedic Surgery* 1921; 3(6):259-67.

[52] Osgood, Backgrounds of orthopaedic surgery, 265.

This recognition implies that the path forward must include more substantial training and standards of surgical technique. Several AOA Presidential Addresses in this period pressed for increased requirements for general surgery training. In an additional step, Osgood reframed the grounding of orthopaedic surgery:

> *We believe the day is near when it will be generally recognized that a new problem in the Surgery of the Extremities and Spinal Column is more likely to be solved by an orthopaedic surgeon than by any other. We believe it is already recognized that an orthopaedic surgeon, whose training is equal to that of a general surgeon of comparable ability, will probably obtain better results in the treatment of fresh, simple and compound fractures than his general surgical colleague.[53]*

This optimistic quote gives a nearly modern definition of orthopaedics. If positioned as surgeons of the extremities and spinal column, and trained thoroughly in surgical technique, then a rationale would exist for enveloping fracture care.

In the interwar period, this articulation of orthopaedic surgery moved to the forefront. The year following Osgood's speech, the mouthpiece of the AOA, formally known as the *Journal of Orthopaedic Surgery*, changed its name to the *Journal of Bone and Joint Surgery*, which it carries to this day.[54] The connection of this renaming to the re-imagining of the field is compelling. Likewise, the AOA would alter its by-laws in line with an anatomical localization of orthopaedics to the extremities and spine.[55] Some professorships in

---

[53] Ibid 266.

[54] Cowell HR. A brief history of the Journal of Bone and Joint Surgery. *Clin Orthop Relat Res* 2000; 374:136-44. While this article suggests that change occurred in 1921, it appears that the new name appeared in 1922. An announcement to this effect was published in: Osgood RB, Soutter R, Low HC, Danforth MS, Brown LT, Wilson PD. Eighteenth report of progress in orthopedic surgery. *Archives of Surgery* 1922; 5:413-47.

[55] Lipscomb PR. Orthopedics, orthopaedic surgery, musculoskeletology, orthopaedics. *J Bone Joint Surg Am* 1975;47:872-6.

this era were even given in Bone and Joint Surgery, rather than in Orthopaedic Surgery.

Orthopaedists in this period also voiced a closer connection to general surgery, dampening the prior rhetoric of separation from the larger field. The reviewer of *Orthopaedic Surgery*, published by Jones and Lovett in 1922, carefully noted, "The book is a contribution to general as well as orthopedic surgery, the latter of which the authors recognize as only a special branch of the former."[56] Some orthopaedists, once proud of their principle of maximizing the end result of function, now denied their special claim to it. A review of an article on the teaching of orthopaedic surgery demonstrates this acquiescence:

> He suggests that so-called orthopedic surgery differs in none of its principles from so-called general surgery. He suggests that orthopedic surgery should not be taught to undergraduates as orthopedic surgery, but that the orthopedic surgeon teach them certain phases and problems of surgery. To the question, What is orthopedic surgery? Allison answers, "The surgery of the extremities and spinal column, which has the reestablishment of function as its guiding principle.[57]

This quote succinctly demonstrates the shift to an articulation of orthopaedics based on anatomical localization and primarily surgical treatment. Allison had served in World War I, but in a short period transformed the view of his work. Far from an isolated opinion, the editors of the reports on progress in orthopaedic surgery voiced agreement:

---

[56] Osgood RB, Soutter R, Low HC, Danforth MS, Bucholz HC, Brown LT, Smith-Petersen MN, Wilson PD. Twentieth report of progress in orthopedic surgery. *Archives of Surgery* 1923; 6:858-908.

[57] Osgood RB, Soutter R, Low HC, Danforth MS, Brown LT, Wilson PD. Seventeenth report of progress in orthopedic surgery. *Archives of Surgery* 1922; 4:693-748. It is commenting on: Allison N. The teaching of orthopaedic surgery. *J Orthop Surg* 1921; 3:448.

> *Ed. Note. - We find ourselves in essential accord with Allison's point of view. As he says, the part cannot be greater than the whole. Orthopedic surgeons must train themselves to the highest points of surgical skill and judgment, in closest contact with general surgery, seeking to contribute to its needs, and happiest when its methods become the property and the practice of the general surgeon and the internist.*[58]

The advance made by increased attention to functional end results was ceded to the medical profession, and surgical skill was promoted as the goal of training.

Ongoing developments in orthopaedics were presented as part of the larger surgical community. "Reports on progress in orthopaedic surgery" changed its place of publication to the *Archives of Surgery* in 1921, rather than to a strictly orthopaedic journal.[59] The authors of these updates began noting that their purpose was to summarize new work in "surgery of the extremities and spine,"[60] illustrating both the new articulation of the field and its more confined role as an offshoot of general surgery.

This conservative posture emerged in response to the unsettled, disputed status of orthopaedics and fracture care in the interwar period. Fresh from success in expanding their reach during World War I, ironically these practitioners argued that the way forward was

---

[58] Seventeenth report of progress in orthopedic surgery, 1922.

[59] Osgood RB, Soutter R, Low HC, Danforth MS, Bucholz CH, Brown LT, Wilson PD. Fifteenth report of progress in orthopedic surgery. *Archives of Surgery* 1921; 3:181-229. It had previously been published in the *Boston Medical and Surgical Journal*, the forerunner of the *New England Journal of Medicine*. Yet the *Journal of Orthopaedic Surgery*, later the *JBJS*, was already well established.

[60] For example, a January 1922 report was based on reviewing 875 articles "on surgery of the extremities and spine." This caveat had not been a part of prior reports. Osgood RB, Soutter R, Low HC, Danforth MS, Bucholz CH, Brown LT, Wilson PD. Sixteenth report of progress in orthopedic surgery. *Archives of Surgery* 1922; 4:200-56.

---

to accept a more modest position in the division of labor. A sustained focus on surgical training, and locating themselves as a branch of general surgery, responded to the critiques of Scudder and others. Orthopaedists essentially argued that their field was a subspecialty of general surgery. After a thorough basic training in general surgery, interested surgeons could undertake additional study in orthopaedics, somewhat akin to modern fellowships. While only modest successes emerged, such as that at the MGH Fracture Service, this rejoinder with general surgery and operative turn reframed orthopaedics as the surgeons of the extremities and spine.

Resolution and Encroachment in World War II

An AOA President, in summarizing the condition of interwar fracture care, recognized the still-marginalized place of orthopaedists: "at best, they shared the management of extremity trauma with the general surgeons... More frequently than not, the efforts of the orthopaedic services were confined to the management of joint fractures, ununited or malunited fractures, and minor fractures which did not involve soft parts." Yet, he continued, "World War II changed all of this."[61]

The opportunity presented by massive casualties of war would again shape the practice of orthopaedics. World War II placed American soldiers in harms way in multiple theaters. In Europe alone, there were 381,350 wounded soldiers, with two-thirds of them estimated to involve the extremities.[62] In this war, however, it was not a transformation of the boundary of the specialty that led to expansion. The established conception of orthopaedists as surgeons of the extremities and spine would not be altered. Instead, it was the

---

[61] Cave EF. Trauma and the orthopaedic surgeon. *J Bone Joint Surg Am* 1961;583.

[62] Cleveland M, ed. *Medical Department United States Army in World War II. Surgery in World War II, Volume 12: Orthopedic Surgery in the European Theater of Operations.* (Washington DC: Office of the Surgeon General, Department of the Army, 1956) vii.

personification of this definition in one man, who happened to rise to the position of Surgeon General, that delivered the care of fractures in its entirety to orthopaedics. The seemingly huge success in fracture care, however, ran alongside an encroachment on rehabilitation. The operative turn in orthopaedics resulted both in a consolidation of authority and a narrowed boundary of the field's scope.

A Prepared Grounding and a Unique Opportunity

In the course of WW2, it was orthopaedic surgeons who cared for acute fractures and dislocations, among other extremity injuries. Histories of the period recognize the unique importance of one man in giving this authority. Dr. Edwin Cave (1896-1976), who would become the first orthopaedist to serve as chief of the MGH Fracture Service after the war, compares the impact of these two conflicts:

> *If the development of orthopaedic surgery was enhanced by World War I, it was expanded tenfold by the experience gained in the War of 1939-45, thanks in no small measure to one of our members, the late Major General Norman T. Kirk, who was at that time the Surgeon General of the Army.*[63]

The official history of orthopaedic surgery in the war concurred: ""It was a stroke of good fortune that, during a large part of the war, the Surgeon General of the Army was Maj Gen Norman T. Kirk, an experienced orthopedic surgeon, who fully understood the needs of this specialty and was most sympathetic to these consultants."[64] While his appointment as Surgeon General was fortuitous, the career development of Dr. Kirk (1888-1960) had prepared him to see

---

[63] Ibid 584.

[64] Mullins WS, ed. *Medical Department, United States Army. Surgery in World War II. Orthopedic Surgery, Zone of the Interior.* (Washington DC: Office of the Surgeon General, Department of the Army, 1970) xiv.

orthopaedists as the authority on fracture care. Up until WW1, Kirk had practiced for several years as a general surgeon. It was only while stationed at Walter Reed General Hospital in 1919 that he became focused in surgery of the bones and joints. He was one of many general surgeons, discussed previously, who took up orthopaedics in the context of that war. Amputee surgery became his clinical focus, and he emerged as a leader in the field with published articles and textbooks.[65] He briefly studied at MGH in 1925, later becoming Chief of the Orthopedic Section at Walter Reed.

Grounded in general surgery, with an operative focus and established experience, Kirk was exactly the sort of orthopaedist that the interwar redefinition meant to promote. He even once held a post as Bone and Joint Surgeon. Though the happenstance of an orthopaedist being Surgeon General illustrates how individuals can shape the process of boundary formation, casting orthopaedists as surgeons of the extremities and spine and their shared involvement in interwar fracture care positioned the field for success. Under Surgeon General Kirk, Senior Consultants in Orthopaedic Surgery were appointed,[66] and one responsibility of this position was to visit wartime hospitals to advise on necessary changes. This power was utilized in part to advocate replacement of general surgeons who had been given control of fracture care in hospitals.[67]

---

[65] His numerous contributions included: Kirk NT. *Amputations: Operative Techniques*. (Washington, DC: Medical Interpreter, 1924). Later editions were published in 1942 and 1943.

[66] While Frank Ober of MGH was a consultant to the Secretary of War, Harvard orthopaedists played a less conspicuous role in WW2. This is itself a reflection of the field's growth. The nucleus of pioneers in Boston had by this point succeeded in broadening orthopaedics, but one result was greater parity in leadership. The strong representation of Harvard in the first decades of AOA Presidents had trended well downward by the time of WW2.

[67] *Zone of the Interior*, 25. In this exhaustive history, an example is given of visiting a hospital where the chief of the orthopedic section was a general surgeon, "without either the training or the experience necessary for this assignment" who was critiqued for after-care, diagnostic procedures, and surgical technique. The Consultant in Orthopedic Surgery recommended multiple chiefs be replaced, as in this case.

A few examples of what constituted an orthopaedic service illustrate their dominance of fracture care. In one hospital center in Europe, between January and June of 1945, there were 42,372 admissions, of which 10,647 were classified as orthopedic. Fractures accounted for more than 5,000 of these cases. Open reduction was only completed for 202 patients, while plaster-of-Paris splints were used for 10,448 and skeletal traction for 2,134. That hospital center had 18,636 beds, and the largest single block (7,500) was assigned to orthopedic surgery.[68] In the 22d General Hospital, between June of 1944 and June 1945, there were 2,460 admissions classified as orthopaedic, of which more than 1,300 were fractures.[69]

Sustained Orthopaedic Authority in Fracture Care,
and the Encroachment of Rehabilitation

This breadth and depth of orthopaedic experience in treating fractures was transferred to civil life. At the MGH, Cave in 1948 became the first orthopaedist to solely lead the Fracture Service.[70] A final defeat of Scudder's vision for an inclusive specialty of Traumatic Surgery was the successful orthopaedic opposition in 1957 to the formation of a Board of Trauma.[71]

Unlike the experience after World War I, this wartime authority had a lasting impact. Fifteen years after Allied victory, in a survey of 83 program directors in orthopaedics, 81 responded that fractures were in the control of orthopaedic surgeons.[72] In 1960 a series entitled "Fracture of the Month" started running in the *Journal of the American Medical Association*, written by Harvard orthopaedists.[73] In an

---

[68] *Orthopedic Surgery in the European Theater of Operations*, 9

[69] Ibid 11.

[70] Brown T. Orthopedic Surgery. *In*: Castleman B, Crockett DC, Sutton SB, eds. *The Massachusetts General Hospital, 1955-1980.* (Boston: Little and Brown, 1983), 280.

[71] Cave, Trauma and the orthopaedic surgeon, 587.

[72] Ibid 585.

[73] One example of this series is: Aufranc OE. Fracture of the month number 13: Subtrochanteric

editorial celebrating its first year, while some collaboration with plastic surgeons were noted, Cave makes no mention of general surgery contributions to the effort.[74] The medical profession now recognized orthopaedic surgeons as the authoritative experts on treating fractures.

Events surrounding World War II provided an opportunity for orthopaedic surgeons to solidify their control over the acute care of fractures. Despite their increased numbers, there was still a shortage of qualified orthopaedists. The demands of acute care, coupled with their focus on operative management, found orthopaedic surgeons during the war becoming increasingly located in general and large station hospitals managing the initial care of injuries.[75] However, the rehabilitation needs of these soldiers were also great.

In the 1910s orthopaedists had expanded largely by advocating for attention ultimate functional outcomes, and stressed adequate after-care and rehabilitation to achieve these goals. In WW2, however, it was the emerging field of physiatry that would organize to care for soldiers after the 'acute phase' of treatment was completed. By outlining these developments, we can see an example of orthopaedics being the target of encroachment, and another result of the operative turn of the interwar period. Shifts in the scope of orthopaedics would secure the acute treatment of fracture, but also risk their prior authority over rehabilitation.

In *The Making of Rehabilitation*, Gritzer and Arluke describe the development of Physical Medicine & Rehabilitation (PM&R) as a specialty.[76] Defining moments in PM&R were also related to the

---

fracture of femur. *JAMA* 1961; 177(4):254-7.

[74] Cave EF. Fracture of the month - one year. *JAMA* 1961; 177(4):258.

[75] *Orthopedic Surgery in the European Theater of Operations*, 16. This transfer of orthopaedic surgeons to general hospitals to meet the demands of the work was pushed by Surgeon General Kirk. *Orthopaedic Surgery in the Zone of the Interior*, 23.

World Wars. Out of WW1 came not only the creation of physical and occupational therapy, but also a professional association of physical therapy physicians. However, rehabilitation was not a central function of these physical therapy physicians until well into World War II. Instead, they attempted to contribute to acute treatment using electrical devices and other treatments such as infrared lamps or therapeutic baths. Unlike orthopaedic surgery, this group of so-called physical therapy physicians had failed up through the 1930s in efforts to be recognized as a medical specialty.

The success that orthopaedic surgery had in advancing their position through the course of World War I was seen as a blueprint for physical therapy physicians in World War II. The influential Baruch Committee on Physical Medicine published a report in early 1944, which argued, "the last war is said to have established orthopedic surgery as a recognized specialty... this war may do the same for physical medicine."[77] To achieve this goal, the committee pressed the need to increase their supply of practitioners, advocate via public relations to increase their recognition, and be of special use for wartime and postwar treatment of soldiers.[78]

To provide useful wartime service, the Baruch Committee argued that these physicians would take over "... the dead space between definitive care and ability to return to productive work, the setup for retraining and reconditioning, medicine and its relation to environment, occupation, social status, etc."[79] The vacuum into which they aimed to enter was left open by orthopaedists who had shifted their focus to the operative treatment of acute injuries.

---

[76] Gritzer G, Arluke S. *The making of rehabilitation: a political economy of medical specialization, 1890-1980.* (Berkeley: University of California Press, 1985).

[77] *Report of the Baruch Committee on Physical Medicine.* (New York: Baruch Committee on Physical Medicine, 1944) 1.

[78] For further discussion of the Baruch Committee, see Gritzer and Arluke, *The making of rehabilitation,* 95-99.

[79] The Baruch Committee on Physical Medicine. *Journal of Rehabilitation* 1945; 11:32.

As part of reconceptualizing their work, physical therapy physicians changed their field's name in 1944 to Physical Medicine, both to "lend more dignity to its practitioners" and to "overcome much of the confusion by the indiscriminate use of the term' physical therapist' by both physicians and technicians."[80] Soon after the war, a resolution was adopted to title individuals working in this field as physiatrists. These changes reflect both the process and product of the field's expansion, which would come largely at the loss of orthopaedic authority over rehabilitation.

As orthopaedics shifted its scope to focus on operative treatment, Goldthwait had cautioned of the inherent risks:

> *If we are to see only the operation, leaving the after-care to the slightly trained house staff with the physiotherapy given by an entirely different department, of which we have little control or knowledge, we cease to be true orthopaedic surgeons, but just surgeons doing bone and joint work.*[81]

As a prior proponent of principle-based orthopaedics, Goldthwait felt this shift could mitigate their claims to rehabilitation. Moreover, he recognized that the operative turn was a fundamental shift in the boundary of orthopaedics.

In the course of the war, this prediction came to pass. While physical therapy had been placed under control of orthopaedic surgery in 1942, by 1946 it was transferred to the newly created Physical Medicine Consultants Division.[82] Physical Medicine's victory came ultimately by 1947, with AMA specialty recognition and the first examination given by the American Board of Physical Medicine.

[80] Gritzer and Arluke, *The Making of Rehabilitation*, 97.

[81] Goldthwait JE. The backgrounds and foregrounds of orthopaedics. *J Bone Joint Surg Am.* 1933; 15:283.

[82] *Orthopaedic Surgery in the Zone of the Interior*, 12.

Soon after, 'rehabilitation' was added to its name. In the following decades attempts were made to by orthopaedic surgeons to stifle physiatrists, which included advocating for a subspecialty of orthopaedic rehabilitation.[83] Although somewhat successful, the necessity of qualifying a particular section of orthopaedics as interested in rehabilitation only further signifies that this prior area of control had become peripheral to the specialty's boundary.

Specialty boundary adaptation in PM&R

The success of PM&R in achieving specialty status and assuming some authority over rehabilitation illustrates the impact of redefining a field's boundaries, as as we have seen for orthopaedic surgery. Gritzer and Arluke realized the importance of such adaptability, writing that for PM&R, "... formal specialty status would be granted not in recognition of their traditionally claimed area of expertise, but for their capacity to shift the definition of their work to meet needs derived from war."[84] By moving from 'acute phase' treatment to expertise in handling the convalescent phase of illness and the disabled, physiatrists both responded to increased need and moved into a domain previously held closely by orthopaedists. The ongoing process of delineation and reimagining is not unique to orthopaedics or PM&R, but a necessary adaptive feature of successful specialties.

Summary

After World War I, the changing epidemiology of illness and an eagerness to find a place in the field of industrial surgery made fracture care a locus of dispute. The development of a Fracture Service at MGH illustrates the halfway legacy of WW1 for

---

[83] This orthopaedic response emerged in the 1960s. For further discussion, see Gritzer and Arluke, *The Making of Rehabilitation*, 148-52.

[84] Ibid 90.

orthopaedic treatment of fractures. While Osgood and others were allowed into the framework of fracture care, general surgeons retained control. Able to use radiographs to diagnosis, treatment, and evaluate outcomes for bony injuries, these two groups of surgeons together developed and propagated advances in knowledge. While expertise was shared, orthopaedic and general surgeons differed in their vision of fracture care. The principal of protecting and restoring function could no longer differentiate orthopaedics, and even if arguments for specialized fracture treatment were accepted, there were multiple conflicting possibilities regarding who should care for these injuries. This lack of clear authority, along with lingering doubts regarding its practitioners' surgical training, resulted in the newly prominent orthopaedic community reframing itself more as a branch of its generalist colleagues. This was the period in which orthopaedists emerged as surgeons of the extremities and spine, poignantly symbolized by its flagship journal changing its name from the *Journal of Orthopaedic Surgery* to the *Journal of Bone & Joint Surgery*.

Recast with a surgical focus, events of the Second World War culminated in orthopaedic control of fracture care at MGH and across the country. This was a victory only if we see fractures as the sole goal of orthopaedists. Rather, it was another boundary adjustment resulting from yet one more rearticulation of the specialty in the context of internal and external developments. As had been feared by prior generations, the shift away from 'mechanical therapies' and a principle of restoring function that was inclusive of rehabilitation left an opening greatly widened by the needs of another wave of injured soldiers. Physical therapy physicians saw WW2 as their own opportunity to expand, and followed the lead of WW1 orthopaedists by redefining their field as physical medicine and expansively embracing chronic rehabilitative care as a core function.

Conclusions and Speculations

Orthopaedic surgery's transformation from its beginnings in the United States through the end of World War II illustrates how a specialty defines its boundaries and then iteratively alters them in response to social, technological, and political forces of change.

Analysis of this narrative has been enriched by giving greater voice to individual physicians as they articulate the meanings and scope of their work in response to continuous develops in the ecology of their practice. Orthopaedic surgeons affiliated with Harvard Medical School played conspicuous roles throughout this period, and their experiences have given substance and structure to this monograph. In addition to adding to our understanding of this tumultuous period for orthopaedics, it provides a lens through which to inform debate about the future of this and other medical specialties.

The word "orthopaedics" has been rearticulated in converging, shifting contexts over time. The techniques of tenotomy, anaesthesia, and antisepsis opened a possibility for surgical treatments of so-called deformities, yet orthopaedics did not suddenly become an operative specialty. While John B. and Buckminster Brown readily supplemented "buckles and straps" with subcutaneous tenotomy, under the same umbrella of orthopaedics Bradford and his protégé Lovett spent their careers using almost entirely non-operative measures like prolonged traction, serial braces, and manipulative therapies. The tension between surgical and mechanical treatment in the early era of orthopaedics, and the divergent practices of pioneers in the field, illustrate but one moment where the path forward was undetermined.

Medical technology is by no means the primer driver of divisions in labor. The opportunity to emerge as a specialty was predicated on increased urbanization, transportation, and a rise of hospitals in which specialized equipment and adequate numbers of patients with similar conditions could come together. A philanthropic focus on crippled children was the background to Bradford's accolades in the treatment of skeletal tuberculosis.

Historical opportunities, such as that provided in World War I, were not automatic gains for orthopaedics. Osgood's exposure in Paris led him and others to expand the boundary of their specialty to include the acute care of injuries, in defiance of a prior generation of leaders who held that orthopaedics should be strictly confined to chronic conditions. This broadened scope was not the result of particular knowledge, but rather a successful campaign to align the field's goals with that of authorities in military and government; namely, the maximum restoration and rehabilitation of soldiers, either to permit a return to active duty or minimize permanent and costly disability. As Osgood acknowledged, it was a lack of such perspective in other fields that presented a chance to redefine their own mission.

In the interwar period, an unsettled condition arose in which general and orthopaedic surgeons both aspired to the treatment of fractures. There were opposing visions for the ideal treatment of fractures, including a proposed field of traumatism that would have a surgical specialty simultaneously care for broken bones alongside any accompanying damage to the body's organs. With dual authority and multiple legitimate possibilities, this situation is similar to the current split interest in spine surgery by orthopaedists and neurosurgeons, as well as hand surgery being undertaken by orthopaedic and plastic surgeons. In the case of fractures, it required both a distinct operative turn, as well as another war, to resolve this dispute in favor of orthopaedists. An inclusive traumatism could not take hold, though recent efforts by general surgeons to operatively repair a greater number of rib fractures suggest that these borders are not entirely closed. Also, as we saw, a surgically-focused orthopaedics lost at least a part of its authority over rehabilitation to physiatrists. It is likely that if a single specialty is to emerge with sole dominance over spine or hand surgery, the posturing involved would similarly leave opportunities for encroachment.

Taken as a whole, this narrative redeems for the individual the possibility of integrating the backgrounds and foregrounds of an era into opportunities to redefine the boundaries of a given field, or else pave new ground altogether. Traditional histories of orthopaedics have detailed the discoveries and discoverers, generally to the neglect of the coexistent forces of society. Others, notably Cooter's *Surgery and Society in Peace and War*, have detailed the political, economic, and social drivers of change. In this work, while recognizing the framing context of each era, we have seen particular practitioners dramatically nudge the course of events at critical junctures characterized by a plurality of options. The opportunity to make such an impact is coupled with a necessity to react.

A prominent orthopaedist published a 2009 editorial entitled "On Rise and Decline."[1] In it, the encroachment of the field's boundaries by physiatry, podiatry, plastic and neurosurgery, as well as primary care sports medicine physicians are lamented. Additionally, recent scandals involving orthopaedic surgeons' close and possibly deleterious ties with industry are seen as signs of a specialty that has lost its path. Increasing subspecialization and a perceived need for fellowship training after residency are suggested as threats to the unity of the field. Meanwhile, the rollout of comprehensive health policy reform threatens to shift payment models that for decades have benefited procedural fields like orthopaedics.

The perspective gained here shed light on these current dilemmas. The surgical-anatomic conception of orthopaedics, which became prominent in the decades after World War I, may increase the risk of fragmentation and encroachment. As discussed, the focus on operative training and treatment provided an opportunity for

---

[1] Sarmiento A. On Rise and Decline. *J Bone Joint Surg Am* 2009; 91:2740-2.

physical therapy physicians to recast their field as physical medicine and lay claim to a neglected rehabilitation as a central function. Beyond rehabilitation, however, if surgical anatomy alone holds orthopaedics together, then legitimate complexity could drive subdivisions of the "extremities and spine." When multiple specialties posses some claim to the spine, hand, or foot, how can orthopaedics justify authority and clearly demarcate their boundaries? In WW1, the principle of maximizing locomotive function briefly supplied this continuity. If orthopaedic surgery is not to become simply a pathway to a subspecialty, it may be time for a rearticulation.

As clamor rises for systems-level change, orthopaedists should recall their field's advocacy for a cohesive care of injured soldiers in the Great War. Instead of fragmentation, could orthopaedics drive a course to comprehensive, collaborative musculoskeletal care? As patients bounce among an increasing array of doctors and allied healthcare professionals, who will be the advocate of mobility? Can comparative-effectiveness research be embraced as end-results studies were, or will the field continue to push for the 'latest-and-greatest' devices of industry over well-proven therapies (whether operative or not)? These questions will be answered by individual practitioners, and the echo of those choices will rise to the field's leadership.

While the editorial's call to a revival of professionalism and a return to the core of orthopaedic practice are appreciated, the narrative and analysis presented here question the very idea of a definable 'soul' to this or any medical specialty. Recall the conception as given by Andry's *Orthopaedia*, and realize how far removed the field has become. Instead, the substantial challenges presented could be seen as changing conditions, be they social, political, technological, economic or otherwise, which signal the need for a response that rearticulates the boundaries of orthopaedic surgery to serve the current requirements of patients and society. The numerous transformations of orthopaedics from its arrival in America through

World War II reveal the labile nature of medical specialization and the ongoing process of boundary development within and between fields.   Individual   practitioners   and   organizations   proposed disparate conceptions of how to respond in each circumstance, and in doing so iteratively redrew the division of labor.

An easily evident lesson from this work is that medical specialties can, and possibly must, be willing to radically change the meanings and boundaries of their work. At critical junctures, individuals have assessed the milieu of social, technological, economic and other forces to frame a re-imagining of their work. This will be required again.

Glossary

AEF	American Expeditionary Forces
AOA	American Orthopaedic Association
HMS	Harvard Medical School
JBJS	Journal of Bone and Joint Surgery
MGH	Massachusetts General Hospital
PM&R	Physical Medicine & Rehabilitation

www.ingramcontent.com/pod-product-compliance
Lightning Source LLC
Chambersburg PA
CBHW022020170526
45157CB00003B/1299